FINISH CARPENTRY

THE BEST OF
Fine Homebuilding

FINISH CARPENTRY

THE BEST OF
Fine Homebuilding

The Taunton Press

Cover photo: Kevin Ireton

**Back-cover photos: Charles Miller (top),
Jeff Kolle (bottom)**

BOOKS & VIDEOS

for fellow enthusiasts

First printing: 1997
Printed in the United States of America

A Fine Homebuilding Book

Fine Homebuilding® is a trademark of The Taunton Press, Inc.,
registered in the U.S. Patent and Trademark Office.

The Taunton Press, Inc.
63 South Main Street
P.O. Box 5506
Newtown, Connecticut 06470-5506

Library of Congress Cataloging-in-Publication Data

Finish carpentry : the best of Fine homebuilding.
        p.   cm.
    "A Fine homebuilding book"–T.p. verso.
    Includes index.
    ISBN 1-56158-183-6
    1. Finish carpentry.   I. Taunton Press.   II. Fine homebuilding.
TH5640.F56   1997
694–dc20   96-41123
                                                                        CIP

CONTENTS

INTRODUCTION

I T'S AXIOMATIC IN BUILDING that whatever trade follows yours in the construction process will cover up the mistakes you leave behind. Hence, the foundation crew assumes the framers will compensate for their basement walls being out of level. The framing crew assures themselves that drywall will hide their bowed studs and out-of-square walls. And the drywall contractor tells anyone who'll listen that "trim will cover it."

Now, I've heard some trim carpenters say "the painter will fix it," but you don't want any of them working on your house. For the most part, the buck stops with the trim carpentry. It's the piping on the cowboy's shirt: everybody's going to see it, and it has to be right.

In this book you'll find articles about trim carpentry collected from back issues of *Fine Homebuilding* magazine. Covering everything from baseboard and crown molding to built-in furniture, these articles were written by experienced trim carpenters–none of whom rely on painters to hide their mistakes.

–Kevin Ireton, editor

Basic Scribing Techniques

A finish carpenter shares his secrets for fitting trim to uneven, unlevel or unplumb surfaces

by Jim Tolpin

The first thing I learned as a finish carpenter was that square corners, plumb walls and level floors and ceilings don't exist on this planet. And because that's just the way it is, it was up to me to learn how to work with these unfortunate divergencies from the way it ought to be. As the finish man, my job was to fit the pretty stuff to the structures that framers and rockers left behind, no matter how crooked they were.

In my quest for perfect fits, I learned how to use bevel squares and base hooks, among other tools, and became proficient in the use of a slightly customized pencil compass. I learned from legendary boat builder Bud MacIntosh how to use something called a spiling batten to solve certain awkward scribing problems, such as fitting the last ceiling board. I even paid homage to the linoleum trade and learned the ingeniously simple "Joe Frogger" method of creating a template that can produce dead-accurate fits every time.

Using the bevel square—A bevel square is a layout tool with a wood, metal or plastic body having an adjustable metal blade attached to one end. The square is used mostly for determining the angle at which a piece of trim needs to be cut to fit tightly against a surface.

My first bevel square came from my grandfather. It's a nice rosewood-bodied job with a 6-in. long blade. It's pretty and has sentimental value, but like many contemporary bevel squares, it's not the best tool for taking angles. This is because its locking lever, which is located at the pivot point of the tool, often sticks beyond the edge of the body and gets in the way. Also, the body is quite thick, which holds the blade away from the stock. This can throw off the angle measurement. What's more, the body is relatively short, which can also produce inaccurate readings.

I like my all-metal Japanese bevel square better (bottom left photo, facing page). It's much thinner than a conventional bevel square; the lock is a knurled knob that's out of the way; and it can be held and locked with one hand.

Although the use of a bevel square may seem straightforward, it's not. Always extend the blade fully before pressing the outside edge of the body against a surface to measure an angle (such as when measuring an inside corner where two walls meet). Any protrusion of the blade beyond the outside edge of the body will hold the body away from the surface it's resting against, throwing off the angle reading.

Also, don't assume that you can simply press the square against converging surfaces to get an accurate reading. Say, for instance, that you want to fit a baseboard to a door casing (top photo, facing page). To measure the angle of the end cut, set the baseboard where you want it on the floor, then place the body of your bevel square on top of the baseboard to measure the angle of the casing. If you simply lay the body of the square on the floor, any bumps or dips in the floor next to the joint will fool the square into measuring a false angle. An alternative is to set a level or a straightedge on the floor and to measure the angle off of that.

Once you measure an angle, be careful not to jar the bevel before you scribe the workpiece. Fortunately, there's an easy way to ensure against the loss of an angle setting on a bevel: Record it with the help of a boat-builder's bevel board.

Saving angles—Boat builders, who confront compound angles on nearly every piece they fit, have developed a simple, shop-made accessory that makes it easy to measure and record a series of angles for future reference at the saw table. Called a bevel board, it's a board with a bunch of lines drawn across it at angles ranging from 0° to 45° (bottom left photo, facing page). The bevel board allows you to measure an angle with your bevel square and then read the degrees of the angle directly from the board. If the angle scale on your bandsaw, table saw or chopsaw is calibrated to the bevel board, you need only to set the saw to the appropriate degree mark and cut away. If more than one angle is being taken at once, the angles are simply recorded on a scrap of wood or paper that represents a story board of the piece or pieces to be cut.

The bevel board should be made of a stable wood, such as mahogany or teak, that has an interlocking, split-resistant grain. You can also use a scrap of ⅜-in. plywood. Using a protractor, scribe the lines to the board with an awl and then fill them in with an indelible ink. Keep the board thin so that it will be lightweight (⅜-in. thick is sufficient); leave room at both ends of the board for indexing the body of the bevel against it; and radius or chamfer the top edge of the board so that you can orient it at a glance. An alternative board, sans the romance (and not quite as easy to read), is made by scratching the lines deeply into a piece of Lexan plastic.

In lieu of a bevel board, you can scribe and label each angle on a wood block right after you

measure it, then reset your bevel square from the block to mark your trim. If you need to quantify an angle in degrees, measure it on the block with a Speed Square.

The base hook—Another homemade tool, called a base hook, eliminates the need for a bevel square in some applications. Similar in concept to a siding gauge (see the cover of *FHB* #47), it's simply an L-shaped piece of a stable, split-resistant wood used primarily for laying out the end cut of baseboard where it butts against standing moldings such as door casings (bottom right photo, facing page). To use the hook, lap it over the baseboard and hold it hard against the standing molding while scribing a cutline across the baseboard. Be sure the faces of your base hook are perfectly square to the edges, or you'll introduce a margin of error.

Scribing to irregular surfaces—Shortly after I became a finish carpenter, I bought a $5 pencil compass like the kind my kids tote in their school bags. It has two adjustable arms, with a metal feeler point at the end of one arm and a pencil at the end of the other (bottom photo, p. 12). For improved accuracy, I heated and bent out the feeler point of my compass slightly so that the point, rather than a portion of its side, contacts the meeting surface. (The meeting surface is whatever is being scribed to; I'll call the piece to be cut the workpiece.)

Although I haven't tried it yet, I recently learned a tip from Gary Katz, a contractor in Encino, California. To ensure that he can always scribe a fine line, Katz fits his compass with a Cross #3503 mechanical pencil (A. T. Cross Co./ATX Marketing, One Albion Road, Lincoln, R. I. 02865; 401-333-1200). This pencil is expensive ($15.50), but it scribes a very fine line, is well made and has a wonderful warranty. No matter how you damage it and regardless of its age, you can return it to Cross, and they'll send you a new one.

When scribing a line with a compass, you are actually transferring the pattern of the meeting surface onto the workpiece. It is very important, as you scribe the line, that the feeler point on one side of the compass not get ahead of or lag behind the pencil point on the other side. Throughout the scribing process, these two points must align parallel to the direction the workpiece will move to contact the meeting surface. If they don't, the result will be an inaccurate pattern and, ultimately, a sloppy fit. (top

Fitting baseboard. Author Jim Tolpin uses either a bevel square (photo above) or a homemade base hook (photo below right) to lay out baseboards for a tight fit against door casings. Bumpy floors can fool a bevel square, so Tolpin always measures the bevel off of a straight-edge or a piece of the baseboard itself. The base hook is used by holding it hard against the casing (or plinth block in this case) while scribing a cutline directly on the baseboard.

Boat builder's bevel board. Etched with 46 labeled lines spaced 1° apart, the homemade bevel board (photo below) makes it easy to read an angle off a bevel square, then adjust a saw to that angle.

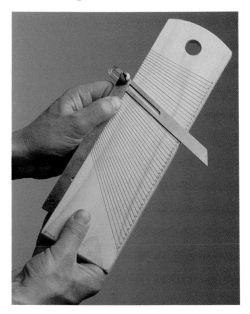

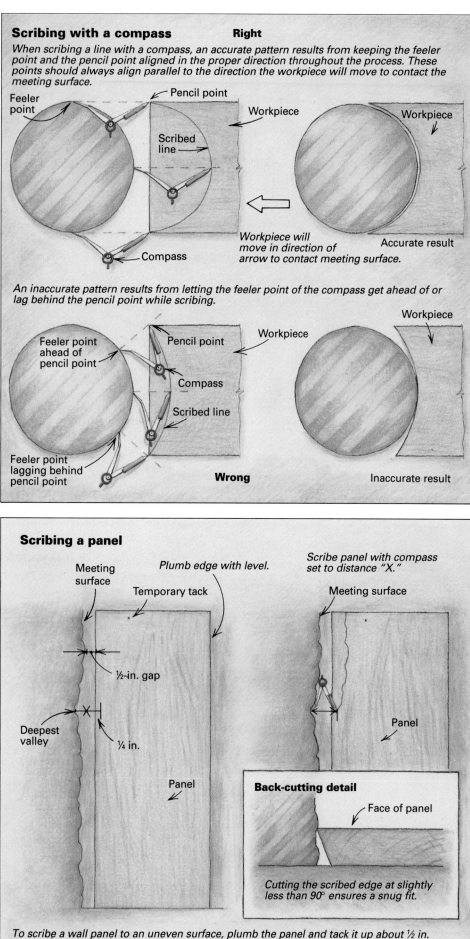

Scribing with a compass   Right

When scribing a line with a compass, an accurate pattern results from keeping the feeler point and the pencil point aligned in the proper direction throughout the process. These points should always align parallel to the direction the workpiece will move to contact the meeting surface.

Feeler point

Pencil point

Workpiece

Workpiece

Scribed line

Compass

Workpiece will move in direction of arrow to contact meeting surface.

Accurate result

An inaccurate pattern results from letting the feeler point of the compass get ahead of or lag behind the pencil point while scribing.

Workpiece

Feeler point ahead of pencil point

Pencil point

Workpiece

Compass

Scribed line

Feeler point lagging behind pencil point

Wrong

Inaccurate result

Scribing a panel

Meeting surface

Plumb edge with level.

Scribe panel with compass set to distance "X."

Temporary tack

Meeting surface

½-in. gap

Deepest valley

¼ in.

Panel

Panel

Back-cutting detail

Face of panel

Cutting the scribed edge at slightly less than 90° ensures a snug fit.

To scribe a wall panel to an uneven surface, plumb the panel and tack it up about ½ in. away from the closest spot on the meeting surface. Then scribe the panel with the compass points set to distance "X" (the distance between the edge of the panel and the bottom of the deepest valley on the meeting surface plus ¼ in.). Back-cutting the panel will ensure a snug fit.

drawing, left). Chinkless-log-home builders, who routinely scribe logs to fit together tightly, have developed a homemade compass with an adjust-able bubble level on it that makes it easy to keep the compass oriented properly while scribing. For more information about this compass, see *FHB* #53, pp. 80-84.

It's amazing what the pencil compass allows you to do. For instance, it really comes into its own for fitting a wall panel or a vertical siding board to a bumpy surface, such as a fireplace (bottom drawing). The procedure is straightforward. First, plumb the panel or board and tack it to the wall about ½ in. away from the closest spot on the meeting surface. Then set the compass to distance "X" between the edge of the panel and the bottom of the deepest valley on the meeting surface, plus ¼ in. so that the scribed line won't fall off the edge of the workpiece. Hold the compass level along the entire vertical run and trace the meeting surface with the feeling point so that the pencil transfers the profile to the workpiece. (If the workpiece is dark, a strip of wide painter's masking tape applied to the panel will make the line more legible.) Finally, remove the workpiece from the wall, back-cut it (bevel it back) a few degrees along the scribed line, then test fit it against the meeting surface. If the fit is good in some areas and way off in others, you probably let the compass wander from level during scribing. If this happens, try again. If necessary, final fitting is achieved through hand planing, sanding, rasping and filing (more on that later).

Sometimes the closing (last) board or panel on a wall must be scribed. This is tougher to do because the board has to fit into an existing gap. For one solution to this problem, see Tom Law's method on the facing page.

Another common scribing problem is fitting stair treads between a pair of skirtboards (or similarly, closet shelves between two walls). This is accomplished by cutting the tread ¾ in. longer than its final length; dropping it into place with one end riding high on a skirtboard; scribing and cutting the low end; marking the final length of the tread by measuring off the scribed end; dropping the tread back into place with its scribed end riding high; and then scribing and cutting the low end to the measurement mark (for an illustration of this trick, see *FHB* #68, p. 61).

Some fitting jobs are accomplished using a pencil compass in concert with a bevel square and a combination square. Laying out a window stool is a good example (photos, p. 12). In this case, use a combination square to locate the outside corners of the window opening (where the wall meets the side jambs); a bevel square to lay out the angles of the side jambs relative to the front edge of the stool; and a compass to scribe the stool horns to the wall. For added convenience, a couple of sticks tacked to the sill (perpendicular to the window) will support the stool while you lay it out.

Scribing with a spiling batten—Sometimes it's awkward to hold a workpiece in position for scribing. A perfect example is scribing the closing board of a wood-strip ceiling. Not only do you have to hold up the board while scribing it,

Scribing the closing board

by Tom Law

The installation of wainscoting and the installation of vertical siding both have the same problem—fitting the closing (or last) board. I approach this problem by nailing up all but the last few boards. Then I tack up the rest of the boards except for the last one. I mark and cut this closing board, remove the tacked ones, then spring the whole group into place at once and nail them to the wall.

How I fit the closing board depends on the nature of the surface it meets. If the meeting surface is straight and plumb, I simply measure the gap and rip the board to width. If the meeting surface is straight but not plumb, I measure across the top and bottom of the gap, transfer the measurements to the board, connect the marks with a straight line and rip the board with a circular saw.

If the meeting surface is irregular, the board needs to be scribed as shown in the drawings below. In this case, after I've tacked up the next-to-last board, I mark the top and bottom of its leading edge on the underlying wall (points A and B). Then I remove all of the tacked-up boards, hold the closing board hard against the meeting surface and mark the top and bottom of its trailing edge on the wall (points C and D).

Setting the closing board aside, I now adjust my compass to span either the top or the bottom two marks, whichever are the farthest apart (points A and C in the example). I then use the compass to make a new mark (point E) at the opposite end so that the top and bottom pairs of marks are the same distance apart (if they weren't to begin with). Finally, I replace the closing board so that its trailing edge falls on the appropriate marks (C and E), then use the compass, with its setting unchanged, to scribe the board to the meeting surface.

I almost always make the cut with a handsaw, cleaning up to the line with a block plane if necessary. A handsaw cuts on the downstroke, ensuring that any tearout will occur on the backside of the board where it won't show. I undercut the board slightly so that when it's sprung into place it makes a nice, neat joint.

When installing wall paneling instead of boards, only the second-to-last panel is tacked up and removed to fit the last piece. Otherwise, the scribing and cutting procedure is the same.

—Tom Law is a consulting editor of Fine Homebuilding *and lives in Westminster, Md.*

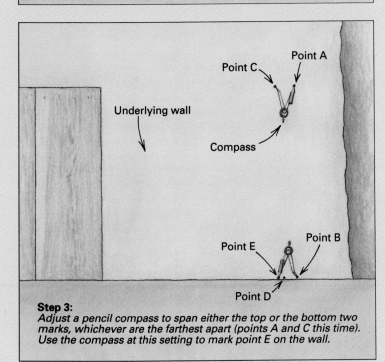

Here's a four-step method for fitting the last vertical board on a wall to a bumpy surface.

Point A

Last few boards tacked temporarily to wall.

Gap to be filled by last board.

Point B — Underlying wall

Step 1:
Install all but the last board on the wall, tacking up the last few boards for easy removal. Mark the leading edge of the second-to-last board on the wall (points A & B).

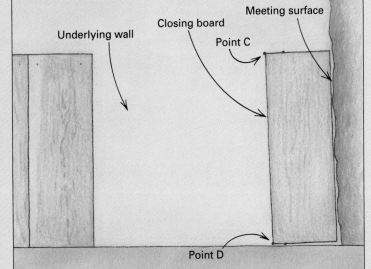

Meeting surface

Closing board

Underlying wall

Point C

Point D

Step 2:
Remove the tacked-up boards, hold the closing board hard against the meeting surface and mark the top and the bottom of the board along its trailing edge (points C & D).

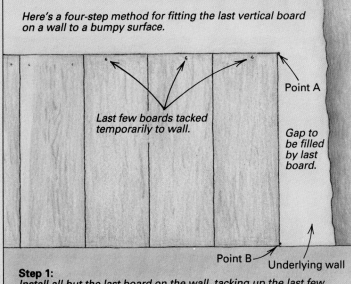

Underlying wall

Point C

Point A

Compass

Point E

Point B

Point D

Step 3:
Adjust a pencil compass to span either the top or the bottom two marks, whichever are the farthest apart (points A and C this time). Use the compass at this setting to mark point E on the wall.

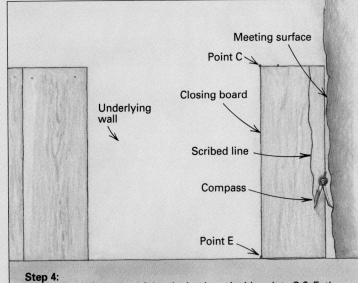

Meeting surface

Point C

Underlying wall

Closing board

Scribed line

Compass

Point E

Step 4:
Align the trailing edge of the closing board with points C & E, then scribe the board off of the meeting surface with the compass setting unchanged. Once the board is cut to fit, spring it and the remaining boards in place and nail them to the wall.

Fitting a window stool. The author uses a combination square to mark the outside corner of the window opening on the stool (top photo above). Then he uses a bevel square to measure the angle between the window jamb and the front edge of the stool. This angle is scribed on the stool through the corner point (middle photo above). Finally, he sets his pencil compass to span the distance between the corner of the opening and the corner mark on the stool, then scribes the stool's horn to the wall (bottom photo above).

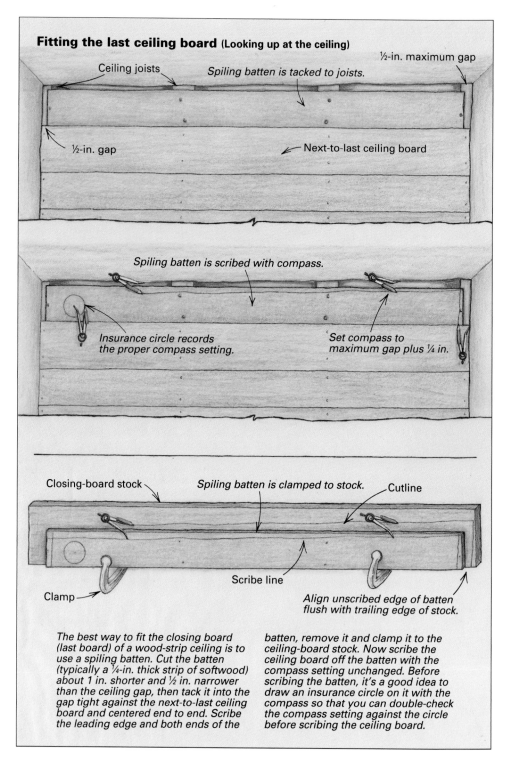

Fitting the last ceiling board (Looking up at the ceiling)

Ceiling joists

Spiling batten is tacked to joists.

½-in. maximum gap

½-in. gap

Next-to-last ceiling board

Spiling batten is scribed with compass.

Insurance circle records the proper compass setting.

Set compass to maximum gap plus ¼ in.

Closing-board stock

Spiling batten is clamped to stock.

Cutline

Clamp

Scribe line

Align unscribed edge of batten flush with trailing edge of stock.

The best way to fit the closing board (last board) of a wood-strip ceiling is to use a spiling batten. Cut the batten (typically a ¼-in. thick strip of softwood) about 1 in. shorter and ½ in. narrower than the ceiling gap, then tack it into the gap tight against the next-to-last ceiling board and centered end to end. Scribe the leading edge and both ends of the batten, remove it and clamp it to the ceiling-board stock. Now scribe the ceiling board off the batten with the compass setting unchanged. Before scribing the batten, it's a good idea to draw an insurance circle on it with the compass so that you can double-check the compass setting against the circle before scribing the ceiling board.

but the oversize board tilts into the opening. This tilt can throw off the scribed line.

Boat builders confront this exact situation when planking a wooden hull, and they've come up with a nifty device to cope with it: the spiling batten. The spiling batten is simply a thin strip of wood (¼-in. thick softwood is standard) that's tacked into the opening that the last plank (or shutter) will have to fill. The batten is scribed (or spiled, as boat builders would say) to the meeting surfaces along its leading edges and ends, then removed and clamped to the workpiece. The scribe is then reproduced in reverse, from the batten back to the work.

For ceilings, I cut the batten about 1-in. shorter and ½-in. narrower than the ceiling gap, then tack it up snug against the next-to-last ceiling board, spaced ½-in. shy of the wall at either end (drawing above). I then set my compass to the maximum gap between the batten and the wall, plus ¼ in. to make sure that the scribed line doesn't veer off the edge of the batten. Before scribing the batten, I draw a circle on the batten with the compass to serve as a reference if the compass is bumped inadvertently. Once the batten is scribed, I remove it and clamp it to the closing board, positioned with its trailing edge (the edge that meets the second-to-last ceiling board) flush with the trailing edge of the board. The board is then scribed off the batten with the compass setting unchanged (double-checked against the insurance circle on the batten).

I back-cut the ceiling board about 5° to allow the board to swing into place. And this cut makes it easy to plane the board to fit if necessary.

Fitting floors to posts—The bottom right photo on the facing page shows a wide floorboard that fits tightly around a post. If the post had been square and its faces flat, I would have laid out the floorboard using a combination square. But, of course, the post isn't perfect, and the combination square stayed in the toolbox.

Instead, I called on the Joe Frogger method, as it's known in the linoleum trade, to make a template that works like a spiling batten. You'll need a pencil, a utility knife, a piece of heavy felt paper or noncorrugated cardboard for the template and a small block of wood measuring about 1½ in. square by about ½ in. thick (the block is the frog).

Joe Frogger. **A trick of the linoleum trade called Joe Frogger makes it easy to fit a floorboard to a post. First, Tolpin cuts a cardboard template to fit around the post and tapes it to the subfloor, tight against the previously installed floorboard. Next, he holds a small block of wood (the *frog*) against the post in several spots while marking the frog's outside edge on the template (photo above). After removing the template, he tapes it over the next floorboard and uses the frog to transfer each mark from the template to the board (photo above right). The marks are joined using a straightedge and a pencil. The reward is a perfect fit (photo right).**

The procedure is simple (photos this page). First, use the utility knife to cut an opening in the template that matches the profile of the post, adding about ¾ in. clearance all around. Slip the template around the post, tight against the last installed floor board, and attach it to the subfloor with double-stick tape. Then hold the frog against the post at stations spaced a couple of inches apart while you mark along the outside edge of the frog on the template with a sharp pencil. Rabbets at opposite ends of the frog make it easy to orient the frog in the same way at each station (scribing is always done off a rabbetted edge).

Next, remove the marked template from the subfloor and tape it to the floorboard to be fit, flush with the board's end and trailing edge. Then index a rabbetted end of the frog against each mark on the template while you mark the opposite end on the floorboard. Finally, join the marks using a pencil and a straightedge, then back-cut slightly along the cutlines. If you're careful, you'll be rewarded with a perfect fit.

Cutting it—Once a workpiece is laid out, there are a number of ways to cut it. Unless the cutlines are relatively straight, allowing the use of a circular saw, I always use a Bosch 1581VS jigsaw to cut just to the line. The saw blows dust off the cutline, its reciprocating blade cuts fast, and its tilting base allows back-cutting. Besides making it easy to trim stock for a tight fit, back-cutting allows the workpiece to be squeezed into place.

I use a block plane and rasps to remove stock up to the cutline, skewing the block plane to reach into dips. Fine-tuning is accomplished with flat and round files.

I've worked with a guy who insists that a belt sander is faster and more controllable than a jigsaw for wasting stock to a wiggly line. Another guy uses an angle grinder. Still another scribes with a small bandsaw, which he outfits with a pair of wheels to make it maneuverable on the job site. □

Jim Tolpin is a finish carpenter, cabinetmaker and writer in Port Townsend, Wash. His book, Working at Woodworking, *is available from The Taunton Press, Inc.; (800) 888-8286. His manual on finish carpentry was published by Craftsman Book Company. Photos by Patrick Cudahy except where noted.*

Hand Planes for Trim Carpentry

Tuned and adjusted right, these planes will save time and improve your work

by Scott Wynn

More than 100 different wood and metal hand planes are described in R. A. Salaman's book *Dictionary of Woodworking Tools: c. 1700-1970* (published by The Taunton Press, Inc.). Store-bought or handmade, many of these clever devices were once indispensable to builders. Before the advent of power planes and routers, a carpenter's repertoire might include assorted bench planes for preparing and smoothing wood stock; molding planes for shaping everything from stair nosings to door casings; and various contraptions for plowing dadoes, grooves and rabbets. A specialized carpenter might even own a compass plane for cutting convex or concave curves and a "galloping jack" plane for smoothing floorboards.

Nowadays, most of these planes are prized more by museum curators and tool collectors than by carpenters. But some types remain as vital on the job site as ever. As an architect/builder who specializes in trim carpentry, I use several kinds, primarily for fitting wood trim or casework against previously installed work, or wherever the use of a power plane or a router is impractical. My favorites are the block plane, the shoulder plane and the butt mortise plane. I also use an assortment of specialty planes (I made some myself) for cutting roundovers and chamfers.

Hand planes are available from woodworker's suppliers, mail-order tool outfits and some hardware stores and lumberyards. But don't expect planes to make smooth cuts straight out of the box. Properly tuned and adjusted, though, they'll cut wood like butter and sing while they work.

The block plane—The typical metal block plane (drawing right) features an adjustable blade housed in a 6-in. to 7-in. long metal body. Mounted bevel-side up, the blade is clamped by a lever cap in two areas: against either one or two milled plateaus or a lateral adjustment lever at the top end of the blade, and against a narrow angled seat at the bottom end. The seat is directly behind the throat (the opening in the sole of the plane through which the blade projects). Depth of cut is controlled by turning a knurled nut or knob at the back of the plane.

Unlike its larger siblings—the jointer plane, the jack plane and the smoothing plane—the block plane is designed for one-handed use and will fit into most tool pouches. These attributes make it the plane of choice for most carpenters. I use mine for trimming miters, fine-tuning the fit of passage doors and flush-mounted cabinet doors,

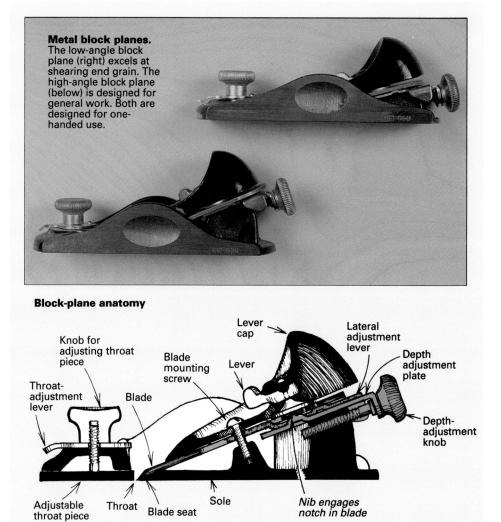

Metal block planes. The low-angle block plane (right) excels at shearing end grain. The high-angle block plane (below) is designed for general work. Both are designed for one-handed use.

Block-plane anatomy

Knob for adjusting throat piece

Throat-adjustment lever

Blade mounting screw

Blade

Lever cap

Lever

Lateral adjustment lever

Depth adjustment plate

Depth-adjustment knob

Adjustable throat piece

Throat

Blade seat

Sole

Nib engages notch in blade

cleaning up jigsaw cuts, fitting cabinets to walls, flush-trimming screw plugs, planing door jambs flush with adjacent walls before installing casings and plenty of other routine tasks.

The planes that most carpenters are familiar with are the Stanley No. 12-020 and No. 12-060 (photo above) and the Record No. 09½ and No. 060½, though similar tools are made by other manufacturers (I own Stanleys). The 12-020 and the 09½ bed the blade at about 20°. The other two bed it at 12°. A low-angle plane is best for planing softwoods, hogging off wood and shearing end grain; a higher-angle plane cuts hardwoods with less tearout. Both types, however, perform so well when properly tuned that it's

hard to tell the difference between them except under the most demanding circumstances.

All four of these planes have an adjustable throat, an important feature for preventing tearout—especially when making exceptionally fine cuts. Some block planes don't have an adjustable throat: Don't bother with them.

Body work—For a block plane to work right, its sole must be flat, its blade properly bedded, and the front edge of its throat must be smooth and parallel to the blade's cutting edge. The blade must be sharpened, with its back flat and free of imperfections (for more on sharpening, see the sidebar on p. 18).

Woodworkers have long debated the wisdom of flattening plane soles. Some argue that planes come flat enough from the factory, but I think a few minutes spent flattening a plane sole can improve performance significantly. Block-plane soles don't *really* have to be flat along their entire length. What matters most is that three areas of the sole contact a flat surface: the throat and both ends. If the throat area is relieved even slightly, the plane performance will be diminished.

Before flattening the sole, retract the blade but don't remove it. This way the plane body is stressed as it would be in use. I flatten the sole by rubbing it on a dry sheet of 600-grit wet-or-dry sandpaper laid on plate glass, being very careful not to rock the plane in the process. You can also use a saw table or a jointer bed instead of glass if you're sure they're flat (they usually aren't). The high areas of the sole will develop a dull, gray color that's easily distinguishable from the low spots. When the throat area and both ends of the sole turn this color, you're done. If at first the throat doesn't touch the sandpaper, I switch to 220-grit sandpaper to speed up the process, then to 320-grit, 400-grit and finally 600-grit paper once the throat makes contact. Finally, I smooth the edges of the sole with a file to remove any burrs or imperfections.

Next, inspect the blade seat to make sure that no burrs or bumps remain from incomplete milling. High spots can be leveled by removing the adjustable throat piece from the plane and flattening the bumps carefully with a fine file. If you don't see any bumps, don't touch the seat: You'll have a tough time restoring it if you mess it up.

Now mount the blade (and the throat piece if you've removed it) in the plane, sight down the sole and adjust the blade so that it protrudes $\frac{1}{32}$ in. or so, with the cutting edge parallel to the sole. Then adjust the throat piece so that it almost touches the cutting edge of the blade. Hold the plane up to a light and sight through the throat to make sure that the cutting edge is parallel to the edge of the throat piece. If it isn't, or if the edge of the throat piece isn't smooth and sharp-edged, remove the throat piece and file it where necessary. The throat edge must be straight and sharp. Do not round the edge of the throat piece, or the edge won't bear effectively on the workpiece to help prevent tearout.

Lastly, if you plan to use the plane with a miter-shooting board (see sidebar, p. 16), use a square to check that the sides of the plane body are relatively flat and are perpendicular to the sole. (If you plan to buy a new plane, check it for square in the store first so that you don't get stuck with a lemon). Carefully file off any high spots on both sides. Now the plane is ready for action.

Using the block plane—To use the block plane, mount the blade bevel up in the body and clamp down the lever cap, making sure that the lever cap's adjusting screw is tight enough to prevent the blade from being pushed around easily (but no tighter or you risk damaging the plane). Next, set the depth of cut and the throat opening according to the work you are doing. Flip the plane over, sight down the sole and adjust the

The rabbet plane. The Stanley No. 12-078 rabbet plane has two blade seats for regular rabbeting (above) or bullnose work. It comes with a cutting spur for cross-grain work, an adjustable fence and a depth gauge. Photo by Vivian Olson.

The 3-in-1 plane. The interchangeable nosepieces of the Record No. 311 "3-in-1" plane allow it to be used as a shoulder plane for rabbeting (below), a bullnose plane for working in confined spaces or a chisel plane for cutting stopped rabbets.

plane so that the entire cutting edge appears at the throat as a black hairline. Now hold up the plane to a light source and adjust the throat piece so that the throat opening (the distance between the throat piece and the blade) is about $\frac{1}{32}$ in. for planing hardwoods or $\frac{3}{64}$ in. for planing softwoods. To combat tearout, the throat opening should be no wider than the thickness of the shaving. Test your settings by taking a few shavings from a wood scrap. For fine work, the shavings should be straight or rippled and thin enough to read through. If the throat jams, the opening is too narrow or the blade is set too deep. Adjust the plane and try again, repeating the process until you get the shavings you want.

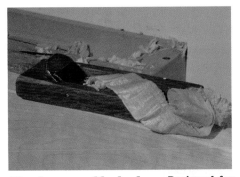

The Japanese block plane. Designed for maximum control, the author's Japanese block plane features a laminated-steel blade that holds an edge longer than western blades do.

Making a miter-shooting board

Trimming small, short pieces of wood with a power miter saw is dangerous. A hand miter box won't trim less than a saw kerf's width (if that), and I'm not ready to buy a miter trimmer, a pricey tool that resembles a paper cutter. The solution to this dilemma is very old: the miter-shooting board, also known as a bench hook (drawings below). It's cheap, portable, safe and is less likely than expensive tools to walk away when your back is turned. Better yet, it can be used with a block plane, which lives in most carpenters' and cabinetmakers' tool kits.

I made my shooting board from scraps. The shooting edge should be made out of a durable material (such as ¼-in. tempered hardboard) glued to a ½-in. to ¾-in. thick plywood base. The miter block should be made out of a 1-in. thick composite material, such as particleboard or medium-density fiberboard (cross-grain movement of a solid-wood block would affect its accuracy). I cut the miter block using a power miter saw, then screw it to the base so that the block can be easily replaced if it's damaged or worn. A hardwood cleat is glued to the base so that the shooting board can be hooked over the edge of a worktable

during use. I also rub a little candle wax to reduce friction where the plane will contact the board.

Before using the board, make sure your plane's sides are square to the sole. If not, file the sides until they are (mine only needed a touch-up in a few spots). Make sure the blade is sharp, and set it for a very fine cut with the throat open wide (tearout isn't a factor when planing across the grain). Then lay the plane on its side on the shooting board, making sure that it rests flat against the base of the jig and the shooting edge. Move the plane to engage the workpiece and then, with one firm stroke, remove a continuous shaving. Don't rock the plane during the stroke. Also, don't get a running start and crash into the piece, and don't chop at it. If you have to chop, either your blade is set too deep or it needs to be sharpened.

I usually take the board right to the area I'm working on so that I don't have to walk around after every stroke or two to check the fit. Also, with a little practice, you can tilt either end of the workpiece off the miter block to trim the piece for out-of-square conditions. —*S. W.*

Block planes will also hog off wood. To do this, open the throat about ⅛ in. wide to prevent overheating of the throat piece and the blade. Adjust the blade downward incrementally until you get shavings of the desired thickness.

To preserve the cutting edge, don't bang it against the workpiece when beginning a cut, and don't drag the plane backwards along the work surface between strokes. Also, I always set down my planes on their sides, not on their soles.

The Japanese plane—Despite their versatility, my metal block planes have one drawback: limited durability of the cutting edge. Nowadays, there are high-quality aftermarket blades available that hold an edge longer than my stock blades do. One company, Hock Handmade Knives (16650 Mitchell Creek Dr., Fort Bragg, Calif. 95437), offers handmade, high-carbon-steel replacement blades for under $20.

Nevertheless, 16 years ago while searching for an alternative to my quick-dulling metal block planes, I bought a small Japanese plane (bottom photo, p. 15). Designed to be pulled instead of pushed, this plane has a 1¾-in. wide, laminated steel blade wedged with a laminated-steel chipbreaker into a 7¼-in. long wood body (roughly the same length as my metal block planes). Though I often use my Stanley planes, I actually prefer using my Japanese plane on the job site. That's because it's lighter (I think of this every time I lift my toolbox), it fits comfortably into my small hands and my hip pocket, and it's surprisingly durable, having survived even a 35-ft. fall off a scaffold. Conversely, a short drop to a hard surface can crack an iron casting, usually at the throat, which renders the plane useless. I've also found that the plane's pull stroke gives me more control than the usual push stroke does (though I prefer pushing the plane when hogging off a lot of wood).

But the biggest reason I like the Japanese plane is edge durability. The secret to this durability is the marriage of a thin, extremely hard layer of high-carbon steel to a thick, strong layer of soft steel. The hard steel provides the cutting edge; the soft steel supports it. The cutting edge on my Japanese plane has actually shaved the very tops off nails (though the nail usually wins). I can use the plane all day, sharpen it that night and be ready for the next day.

Like their metal counterparts, Japanese planes must be tuned before use. The principles are similar—the blade must be sharpened and bedded properly, and the bottom must be flat—but the execution is a bit trickier. One excellent source of information on conditioning these planes is *Japanese Woodworking Tools: Their Tradition, Spirit and Use* by Toshio Odate (published by The Taunton Press, Inc.). Another wellspring of information on tuning and using hand planes, including Japanese ones, is *The Best of Fine Woodworking: Bench Tools* (also published by The Taunton Press, Inc.).

Japanese planes like mine cost about $45, comparable to the cost of metal block planes. They're available from a number of suppliers, including Hida Tool and Hardware Company, Inc. (1333 San Pablo Ave., Berkeley, Calif. 94702; 800-443-

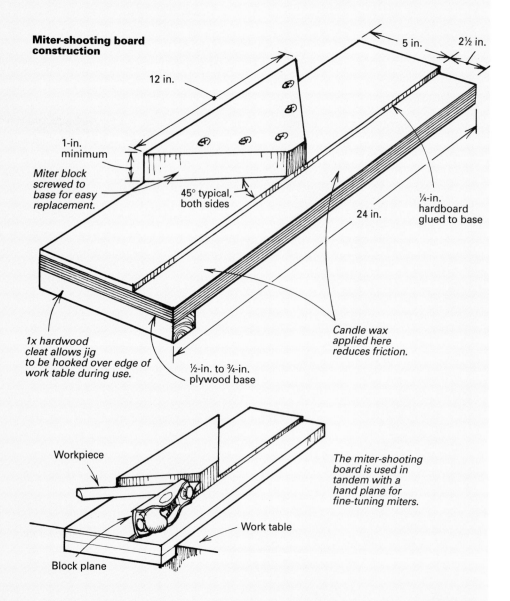

Miter-shooting board construction

12 in.

5 in.

2½ in.

1-in. minimum

Miter block screwed to base for easy replacement.

45° typical, both sides

24 in.

¼-in. hardboard glued to base

1x hardwood cleat allows jig to be hooked over edge of work table during use.

½-in. to ¾-in. plywood base

Candle wax applied here reduces friction.

Workpiece

The miter-shooting board is used in tandem with a hand plane for fine-tuning miters.

Work table

Block plane

Drawings: Scott Wynn

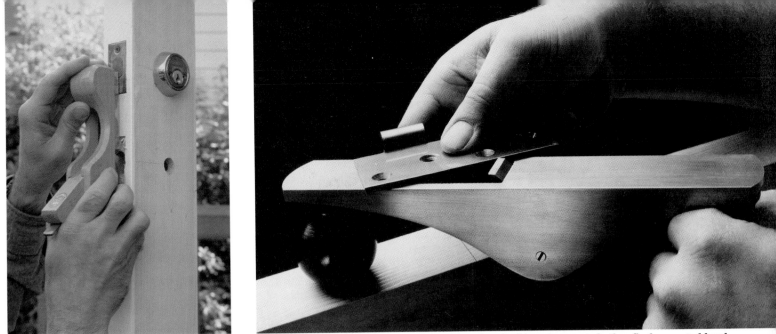

Butt mortise planes. Author Scott Wynn's butt mortise plane (photo above left) cuts level mortises for hinge leaves and other flush-mounted hardware. Lie-Nielsen Toolworks makes the more common metal version (photo above right). Right photo courtesy of manufacturer.

5512) and The Japan Woodworker (1731 Clement Ave., Alameda, Calif. 94501; 800-537-7820).

The shoulder plane—Block planes are the workhorses of trim carpentry, but a few other planes are worth having. I carry a shoulder plane for trimming rabbets because, although I rarely use it, sometimes nothing else will do. This is especially true when fitting new work to old. Rabbets are easily cut with a router or a table saw. But all too often new work is plumb or square, and the old work is not, so the rabbet needs a custom taper. This is easily accomplished using a shoulder plane.

My shoulder plane is an old Record No. 311 "3-in-1" plane (middle photo, p. 15). The 3-in-1 designation refers to three configurations accomplished through the use of interchangeable nosepieces. This allows me to install a long nose for shooting straight rabbets, a short nose for bullnose work in restricted areas, or to remove the nose altogether for chisel-planing to the end of stopped rabbets (rabbets that dead-end).

The 3-in-1 plane has become expensive since I bought mine and now costs about $150 (Clifton makes a similarly priced model). If I had to choose an alternative, I'd pick a Record No. 778 or a Stanley No. 12-078 rabbet plane (top photo, p. 15), which sells for about $65 to $75. Though it doesn't have a chisel-plane mode, it has a bullnose mode and a standard rabbeting mode. It also has a cross-grain cutting spur, a depth gauge and an adjustable fence, making it probably more versatile than the 3-in-1 plane. However, it's too large to fit easily into a toolbox, and it usually takes two hands to use, requiring the use of some clamping system to hold down the work.

Shoulder planes are generally machined more accurately than most brands of block planes, so unless you're having performance problems I wouldn't attempt to tune them. Trying to level the sole on a shoulder plane may tilt the sole out of square with the sides, which is a hassle to correct. Likewise, unless the blade obviously does not sit flat, I wouldn't touch the blade seat.

Chamfering and rounding over. The author's collection of chamfer planes and rounding-over planes includes from left to right: a Radi Plane, which cuts roundovers having radii ranging from $\frac{1}{16}$ in. to $\frac{1}{4}$ in.; a $\frac{1}{8}$-in. radius rounding-over plane; a 1-in. radius rounding-over plane; and an adjustable chamfer plane.

With these types of planes, it's especially important to sharpen the edge of the blade square with the sides because there is virtually no allowance for lateral adjustment of the blades to compensate for an out-of-square cutting edge. The blade should protrude $\frac{1}{64}$ in. or slightly less from either side of the plane body. Otherwise the plane will slowly step out from the shoulder of the rabbet as you plane.

The butt mortise plane—The butt mortise plane is used to cut level mortises for letting in hardware, such as hinge leaves, strike plates and dead bolts. Costing $45, the metal version made by Lie-Nielsen Toolworks, Inc. (Route 1, Warren, Maine 04864; 207-273-2520) resembles a standard plane except that it has a handle at each end, and its throat is wide open (top right photo). I

own a rather obscure German wood model (top left photo) that I bought for $10 at a closeout sale. Like a metal plane, its long throat lets the chips pass through and allows you to watch what you're doing. Given the rather rough nature of mortising, there is no need to tune these planes beyond sharpening the blade.

When using my plane, I first lay out the mortise by outlining it with a chisel, then I make successive cuts with the chisel to the approximate depth required. At this point, the chips would normally be cleared out and the mortise leveled with the chisel. But I use the mortising plane. The blade depth is set to the thickness of the hardware (top right photo) and then the plane is pushed over the chisel cuts, popping out the chips much like a router plane. Then the plane blade is passed over the whole mortise again to remove any high spots. Lastly, the edges of the mortise are squared with the chisel.

The narrow body of this plane allows it to reach confined areas, such as mortises for hinges or strike plates in installed jambs with rabbeted stops. A router is certainly faster for production work, but if you have to cut a variety of mortises, hang a door in an existing opening or install a dead bolt in an existing door, the mortising plane will help speed things up.

Specialty planes—I carry other planes that can be time-savers (bottom photo, p. 17). I have an adjustable Japanese chamfer plane; a rounding-over tool called a Radi Plane (which cuts round-overs having radii ranging from ¹⁄₁₆ in. to ¼ in.); a small Japanese-style, ⅛-in. radius, rounding-over plane that gives an exceptionally smooth finish; and a similar 1-in. radius rounding-over plane.

The chamfer plane allows me to match the chamfers that I machine in my shop on the edges of deck parts or trim, a boon if I need to produce an extra part on site. The Radi Plane and the small roundover plane duplicate the roundovers produced by some of my router bits, as well as those found on a variety of common moldings. The planes also allow me to match the slightly rounded edges typically found on flat stock. Like the chamfer plane, these planes allow me to avoid fussing with a router when I have to produce a simple edge detail or an extra piece of trim. The 1-in. roundover plane is pretty versatile in shaping a variety of radii that you might find on, say, stair nosings or door casings.

The Radi Plane costs about $22. Adjustable chamfer planes cost about $50. The rest are usually priced somewhere in between. ☐

Scott Wynn is an architect/contractor in San Francisco, Calif. He also designs and builds furniture. Photos and drawings by the author except where noted.

Sharpening plane blades

I know carpenters who hone their edge tools by rubbing them on two or three progressively finer sheets of soppy wet/dry sandpaper (ranging from 240 grit to 600 grit), taped or tacked to a scrap of plywood. I've heard of others who sharpen on their belt sanders. My sharpening system is more sophisticated than either of these methods, costs more and takes some time to master, but it produces a superb, long-lasting cutting edge that allows me to do top-of-the-line finish work.

Whatever sharpening system you use, I strongly discourage the use of honing guides. Feel the blade resting on its bevel and develop the body mechanics necessary to maintain that angle while sharpening. You may get frustrated at first, but you'll soon learn to get an adequate edge. As your woodworking skills improve, your sharpening skills will, too.

Sharpening stones—When it comes to producing a sharp, durable cutting edge with a minimum of effort, sharpening stones beat sandpaper every time. The selection of sharpening stones on the market is overwhelming. Oilstones have for generations been the mainstay in the West. Recently developed ceramic and diamond stones promise to combine many of the attributes of other types of stones with few of the drawbacks. For serious sharpening, though, I use Japanese water stones. Though they wear faster than other types of stones and must be flattened frequently, they cut very fast and produce an incomparable cutting edge.

Both synthetic and natural water stones are available. Synthetic water stones are less expensive and less fragile than natural water stones, but good-quality natural stones produce sharper and longer-lasting edges than the synthetic ones do. I use synthetic 1200-grit and 6000-grit water stones to sharpen American and European plane blades. For Japanese blades, I use the 1200-grit stone, an intermediate natural stone called an "Aoto Toishi" (or blue stone) and a deluxe 8000-grit synthetic finishing stone. On the job site, I use a diamond stone to touch up all my blades so that I don't have to deal with water.

Whatever stones you use, buy the best that you can afford. This is especially important for finishing stones, where price does equal quality. My water stones range in price from about $15 for the course stones to $50 for the fine ones. My fine-grit diamond stone cost $56. Most fine-woodworking suppliers carry a full line of sharpening equipment, including Japanese water stones. I got mine from The Japan Woodworker and Hida Tools (see addresses in text).

Grinders—Though you can get by without a grinder, if you use edge tools a lot you'll eventually want one. The two most common types are the bench grinder and the water-stone grinder. Bench grinders work okay, but you have to be very careful with them or they'll overheat the blade and draw its temper, destroying the blade's ability to hold a cutting edge. Bench grinders also hollow-grind the cutting edge, which leaves less metal on the blade than a flat bevel does for supporting the cutting edge. Japanese blades need a flat bevel because their hard, brittle steel at the edge requires the support of the softer, shock-absorbing steel laminated to it. However, any cutting edge subjected to hard use will benefit from a flat bevel.

The water-stone grinder (photo below) overcomes all of the shortcomings of the bench grinder. Water-cooled, it never overheats the blade. And because the revolving water stone is flat, you don't end up with a hollow grind. The water-stone grinder is ideal for beveling nicked or damaged blades. Its only drawback is that the rotating water stone wears out of flat quickly, requiring frequent trueing (I do this using my diamond stone). Water-stone grinders cost up to $300, but the ones I've seen will handle everything from chisels to planer knives.

The work area—Successful sharpening also depends on the nature of the tool-sharpening station itself. If you sharpen at a workbench, the stone should be 4 in. to 5 in. below your belly button. Unfortunately, the typical 3-ft. high bench is much too high for the average person. If the stone is too high, your wrist and elbows will be overly bent, and you'll have trouble maintaining a constant bevel on a blade. Also, your arms will do all the work without any help from your body weight.

I was taught to sharpen on the floor. Kneeling on a pad is pretty comfortable and often brings respite to a back tired from standing for long periods. The floor may be your only alternative on the job site, anyway. If you sharpen on the floor, elevate the stones about 6 in. Mine sit on a homemade redwood water trough (bottom photos, facing page), but you can also use a scrap of 6x6.

Whatever surface you work on, mount stops on it so that the stones don't move around during use, or use one of the manufactured systems that hold and store stones. You can keep synthetic water stones in a lidded plastic tub filled with water so that they'll be ready to go. Don't, however, store natural water stones in water or they'll disintegrate. They also may crack when frozen, even if they're dry.

Flattening the back—The first step in sharpening a new blade is to flatten and polish the back. Don't worry about polishing the entire back, however, just at minimum a narrow flat along the cutting edge (top left photo, facing page). I usually accomplish

The water-stone grinder. Water-stone grinders hone chipped blades quickly. An attached reservoir continuously dribbles water onto the rotating stone to eliminate the risk of overheating the blade.

this using a steel lapping plate and silicon-carbide abrasive powders.

To use a lapping plate, pour ¼ teaspoon of 220-grit abrasive powder on the center of the plate and moisten the powder with a few drops of water (photo 1 below). Lay the blade backside down on the plate, perpendicular to the length of the steel, and rub the blade back and forth. Try to work all of the powder, including the piles that form at each end of the plate. The powder will eventually get very dry and fine, and the high spots on the back of the blade will start to get shiny (as opposed to the dull gray finish elsewhere). Continue rubbing until all of the silicon carbide is a fine paste (you may have to add a few drops of water now and then) and the back has a mirror polish along the entire cutting edge. Maintain even pressure at all times, and be careful not to lift the blade and round the edge. Backing up the blade with a stick helps (photo 2 below).

If the back of the blade is reasonably flat to begin with, I substitute a diamond stone, a 1200-grit water stone and a 6000-grit water stone for the lapping plate. The water stones must be dead flat, though. If a gray oval or large dot appears in the center of the stone while rubbing the blade on it, the stone needs flattening.

On the bevel—Now sharpen the bevel. Soak all but the finish water stones in advance until they stop bubbling (this takes just a few minutes). Sprinkle just enough water on the finish stones to create a slurry during sharpening.

To sharpen, grip the blade between the thumb and forefinger of the right hand (if you are right-handed), wrapping the other three fingers underneath the blade for support (photo 3 below). Holding the bevel flat on the 1200-grit water stone, press down on the edge of the blade with one or two fingers of the left hand and move the blade up and down the full length of the stone, gradually working from left to right and back as you stroke. Ideally, the cutting edge should be perpendicular to the length of the stone; in practice, it's easier to hold the edge diagonally. Keep the stone wet but not flooded. As you stroke, bend your arms and wrists to maintain the blade at the proper angle.

Check your progress by holding up the bevel to a light. The honed portion will be shinier than the unhoned portion. Also, check for a burr by brushing your finger away from the cutting edge. Once the bevel reflects light evenly (photo far right) and you can feel a burr along the entire width of the blade, move on to the blue stone (if you're using one) or to your finish stone. Don't exert as much pressure on these stones as you did on the 1200-grit stone; they polish more with the slurry formed than by direct contact with the stone. On the finish stone, back off (remove) the burr by laying the blade flat on the stone and rubbing it back and forth (photo 4 below). Then flip the blade over and polish the bevel. Alternate between the bevel and the back, shortening the number of strokes per turn until you finish with two or three light strokes on each side. There's no need to polish the edge further with a strop or a buffer.

A 30° bevel works best for most planing. This angle is easy to gauge: The length of the bevel is twice the thickness of the blade. If you're planing softwoods, a 25° bevel will cut cleaner and easier. Some people like to hone a secondary 5° microbevel on the cutting edge. I think this is self-defeating because the microbevel increases friction at the cutting edge and shortens its life. Besides, after the second or third sharpening, a microbevel becomes a macrobevel that requires nearly as much effort to sharpen as a full bevel.

Try to create a convex curve across the width of the blade while honing. This feathers the cut, eliminating steps or ridges across the surface. The curvature of the edge should be virtually indiscernable, equaling the thickness of the shaving you expect the plane to make. This way the blade will cut across its full width for maximum efficiency. An easy way to achieve curvature is by alternately applying pressure on one corner of the blade and then the other while sharpening.

While sharpening, check your water stones from time to time to make sure they're flat. One way to flatten them is to rub them on wet 220-grit or 320-grit wet-or-dry sandpaper laid on a flat surface (such as plate glass laid on a jointer table so that the glass doesn't flex). This technique tends to glaze the stones, however, reducing the cutting action until the top particles are worn away. I prefer to flatten my water stones with a diamond stone; it's quick and doesn't glaze the surface.

Before using the water-stone grinder, I saturate it with water. I don't use the bevel guide on the grinder. Instead, I simply feel the bevel, grinding perpendicular to the edge and moving the blade from side to side to wear the water stone evenly. Be careful that the grinder doesn't grab the blade and throw it, particularly when you first set the blade down. I don't use this grinder to flatten the backs of blades because it grinds too fast and may gouge the blade. —*S. W.*

A sharp blade. The back of the well-tuned blade (above, left) is flat and polished along its entire cutting edge. The bevel (above, right) is honed to a mirror finish. Photos by Bruce Greenlaw.

1. Preparing a lapping plate.

2. Flattening the blade back.

3. Honing the bevel.

4. Backing off.

Plate Joinery on the Job Site

Quick and easy insurance against joints opening up

by Kevin Ireton

Like most people when they first buy a plate joiner, David Mader, a carpenter in Yellow Springs, Ohio, wanted to find out how strong plate joints really are. Mader crosscut a 2x4 and reassembled it with a pair of no. 20 plates (the largest size available), one over the other. After letting the glue dry, Mader tried to break the 2x4 over his knee. He couldn't do it. Convinced of plate joinery's strength, Mader proceeded to use his plate joiner to butt-join custom flooring that wasn't end-matched.

Often considered the province of shop-bound woodworkers, plate joinery, it turns out, is being used more and more by carpenters on the job site (photos at right). Plate joinery and biscuit joinery are the same thing, and in this article I'll use the terms interchangeably.

The basic idea behind plate joinery is simple: plunge a 4-in. circular sawblade into a piece of wood, and you get a crescent-shaped slot. Make a series of these slots along the edges of two boards that you want to join. Insert glue and a football-shaped wooden spline into each slot on one board. Insert more glue into each slot on the other board, then press the two boards together. Water from the glue causes the splines to swell, making a strong, tight joint.

Biscuit-joiner basics—The typical biscuit joiner is a cylindrical machine (drawing facing page), about 10 in. long, and weighs between 6 lb. and 7 lb. It has a D-shaped handle on top and a spring-loaded faceplate in front with an adjustable fence. When the tool is pressed against the workpiece, a 4-in. carbide-tipped blade extends through a slot in the faceplate and scoops out a kerf in the workpiece. You can adjust the distance between the kerf and the face of the workpiece, but any closer than $\frac{3}{16}$ in., and the biscuit, or plate, may pucker the surface of the wood when it swells. You can also adjust the depth of the kerf to fit the size of biscuit you're using.

Biscuits come in three basic sizes (all three are arcs of same circle): #0 is about $\frac{5}{8}$ in. wide and $1\frac{3}{4}$ in. long, #10 is $\frac{3}{4}$ in. wide and $2\frac{1}{8}$ in. long, and #20 is 1 in. wide and $2\frac{1}{2}$ in. long. Biscuits are made of beech with the grain oriented diagonally to the length, making them very strong across their width. Biscuits are also compressed so they'll fit easily in the kerf and then swell once the glue hits them. All biscuits are slightly shorter than the

Slotting in place. To deal safely with a small piece, this carpenter installed the plinth block first and slotted it in place (above). In the photos below, he adds a biscuit, then the casing.

kerf they fit into, which not only allows room for excess glue but also provides some play for aligning a joint along its length. This gives biscuit joinery a distinct advantage over doweling as an indexing technique.

Plate joinery works in hardwood, softwood, plywood, particleboard and even in solid-surface countertops (using Lamello's clear plastic C-20 biscuits). Plates can be used in edge-to-edge joints, butt joints and miter joints.

Joinery comes to the job site—Over the past 15 years, plate joinery has proven itself strong enough and accurate enough to earn a place in many woodworking shops, where it competes with other joinery methods such as doweling, splining or mortise-and-tenon joinery. The merits of plate joinery relative to these other methods can and have been debated. But most carpenters in the field don't enjoy the luxury of a fully equipped shop, and often their only joinery options are whether to use nails or screws. That's why plate joinery adds a valuable technique to a carpenter's repertoire. After all, a cabinetmaker can successfully argue that a biscuit-joined face frame is not as strong as one joined with mortises and tenons. But no one will argue that adding a biscuit between two pieces of mitered casing (photos, p. 22) won't strengthen the joint or greatly improve its chances of weathering changes in humidity without opening up.

Joint strength isn't the whole story, though. A biscuit joiner is very portable, taking up less room in a toolbox than a circular saw. And it's extremely fast. Cutting slots and adding biscuits to a mitered door casing might require 30 seconds. Admittedly, even that little time can be significant when multiplied by a houseful of doors and windows. You might consider it worthwhile, though, if you've ever been disappointed when returning to a job to discover gaps in joints that fit perfectly when you nailed them up.

Who's using them where?—Stephen Sewall, a builder in Portland, Maine, feels so strongly about the advantages of biscuit-joined trim that he seldom installs trim without biscuits. On a recent job where he didn't have his biscuit joiner, Sewall nailed up the side casings, but left the head casings loose so that he could add the biscuits later.

Sewall also says that biscuit joinery has

Drawings: Bob Goodfellow

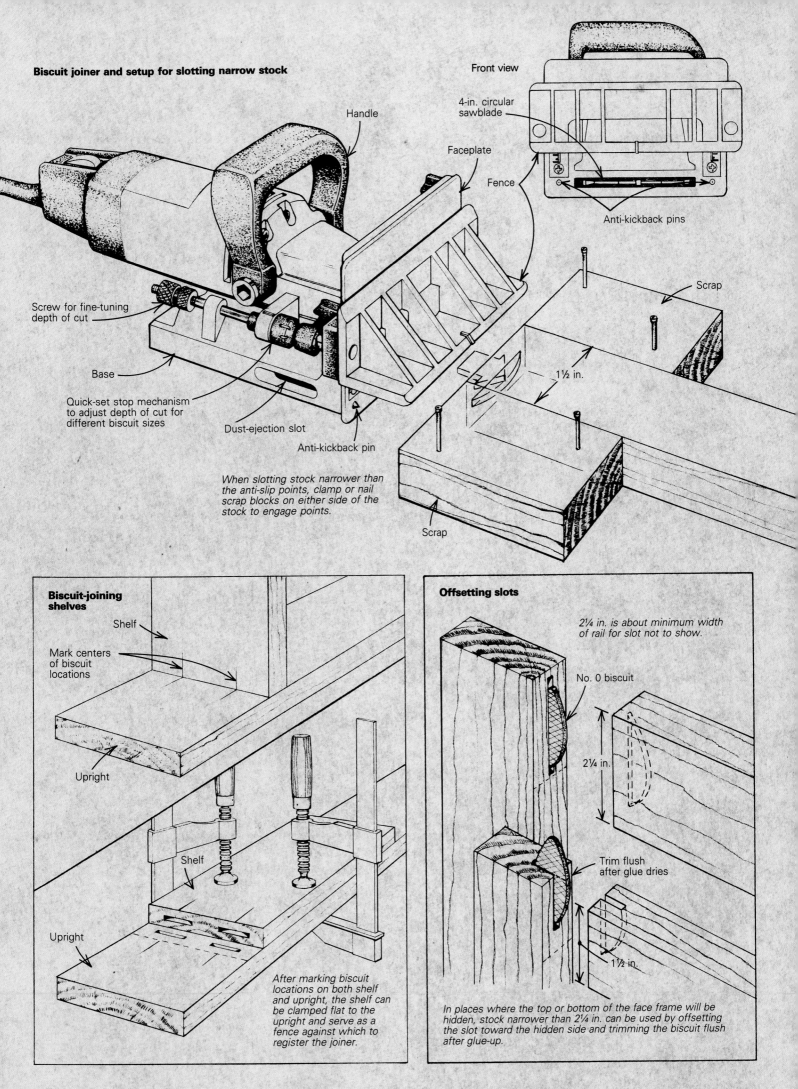

Biscuit joiner and setup for slotting narrow stock

Handle

Screw for fine-tuning depth of cut

Base

Quick-set stop mechanism to adjust depth of cut for different biscuit sizes

Dust-ejection slot

Anti-kickback pin

When slotting stock narrower than the anti-slip points, clamp or nail scrap blocks on either side of the stock to engage points.

Front view

4-in. circular sawblade

Faceplate

Fence

Anti-kickback pins

Scrap

1½ in.

Scrap

Biscuit-joining shelves

Shelf

Mark centers of biscuit locations

Upright

Shelf

Shelf

Upright

After marking biscuit locations on both shelf and upright, the shelf can be clamped flat to the upright and serve as a fence against which to register the joiner.

Offsetting slots

2¼ in. is about minimum width of rail for slot not to show.

No. 0 biscuit

2¼ in.

Trim flush after glue dries

1½ in.

In places where the top or bottom of the face frame will be hidden, stock narrower than 2¼ in. can be used by offsetting the slot toward the hidden side and trimming the biscuit flush after glue-up.

It only takes a second. With the tool and the trim registering against the floor, this carpenter makes short work of cutting slots. By then adding a biscuit spline between two pieces of mitered casing, he prevents the joint from opening as a result of wood shrinkage.

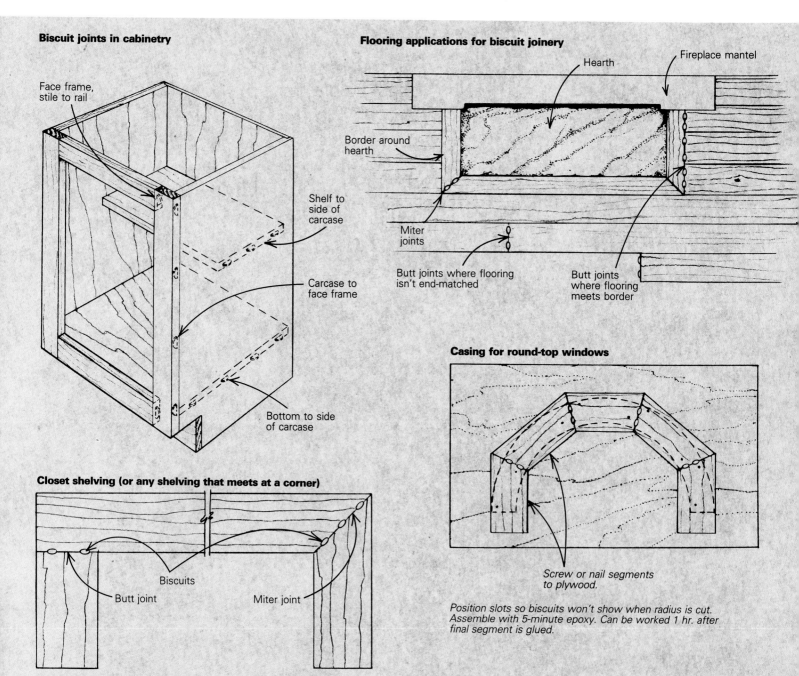

Biscuit joints in cabinetry

Face frame, stile to rail

Shelf to side of carcase

Carcase to face frame

Bottom to side of carcase

Flooring applications for biscuit joinery

Hearth

Fireplace mantel

Border around hearth

Miter joints

Butt joints where flooring isn't end-matched

Butt joints where flooring meets border

Closet shelving (or any shelving that meets at a corner)

Biscuits

Butt joint

Miter joint

Casing for round-top windows

Screw or nail segments to plywood.

Position slots so biscuits won't show when radius is cut. Assemble with 5-minute epoxy. Can be worked 1 hr. after final segment is glued.

made building cabinets on site a lot easier (drawing facing page). When installing a fixed shelf in a cabinet, which he used to do by routing dados in the sides to house the ends of the shelf, Sewall can now biscuit-join the shelf to the carcase faster than he can change bits in his router. When biscuit joining shelves, Sewall often clamps the shelf flat against the upright and registers the biscuit joiner against it as he cuts the slots (see drawing, p. 21). Biscuits will also work in ½-in. stock, like the ½-in. Baltic-birch plywood that Sewall uses to make cabinet drawers.

Foster Jones, a partner in Maine Coast Builders of York, Maine, admits that using biscuit joinery adds to the cost of a job and says his company usually decides before starting a project whether to use biscuits. If they do use them, though, they don't just use them on miter joints. They use biscuits to join inside and outside corners of baseboard, to join baseboard to door casing and to join door casing to plinth blocks (photos, p. 20).

When laying a hardwood floor, the carpenters at Maine Coast Builders use biscuits to join the picture-frame border around a fireplace hearth (drawing facing page) and to join the border to the flooring that butts into it. They even use the biscuit joiner as a trim saw to trim the bottoms of door casing so that flooring will fit beneath it.

Jones also uses biscuits when fabricating trim for round-top windows. Using biscuits and five-minute epoxy, he joins mitered segments of straight stock end to end in a rough semicircle (drawing facing page). He usually screws the segments to a piece of plywood rather than clamping them. The five-minute epoxy lets Jones work with the piece after less than an hour of drying time.

Because biscuit joinery relies in part on the biscuits' capacity to absorb water from the glue and swell up to form a tight joint, you may be wondering how well the system works with epoxy, which isn't a water-based glue. Jones and Sewall wondered, too, and broke apart joints that they had assembled with epoxy. Both found the joint to be just as strong as those made with yellow glue, which has a water base. In fact, Bob Jardinico at Colonial Saw, sole U. S. distributor for Lamello joiners (see the sidebar at right for address), recommends epoxy for biscuit joinery used outdoors. Makes you wonder if biscuits and epoxy aren't the way to keep mitered handrails on exterior decks from opening up.

Plate-joining face frames—A common complaint when assembling face frames (or cabinet doors) with biscuit joints is that the rail must be wider than the slot for the smallest biscuit. Otherwise the biscuit will show. This limits you to a rail that's at least 2¼ in. wide. Responding to this complaint, Lamello recently introduced face-frame biscuits (H-9) that are ½ in. wide by 1½ in. long. But such a short biscuit means you have to switch to a 3-in. sawblade (also available from Lamello).

A company called Woodhaven (5323 West Kimberly Rd., Davenport, Iowa 52806; 319-391-2386) makes biscuits out of particleboard that are 1-in. wide by 1⁵⁄₁₆ in. long and for which you cut kerfs with a 6mm slot-cutting router bit (available from Woodhaven and from MLCS, Ltd, P. O. Box 4053, Rydal, Pa. 19046; 800-533-9298). With these same bits you can also use your router to perform conventional plate joinery, but the cutter is exposed and you don't have a faceplate, so you lose some of the safety and convenience of a plate joiner.

As it turns out, though, you can often get away with using a standard biscuit in a narrow rail by offsetting the slot to the outside of the frame and trimming the biscuit flush (drawing, p. 21). In most cases the exposed kerf and biscuit are either pointing down at the floor and can't be seen, or are covered by a countertop. When cutting slots in narrow stock, it's best to clamp or nail scrap blocks on either side so the steel points on the joiner's faceplate have something to grip (drawing, p. 21). These points keep the tool from slipping during a cut (some joiners employ rubber bumpers or pads rather than steel points).

But wait, there's more—It's easy to think of other job-site uses for biscuit joinery: mitered ceiling beams, jamb extensions on doors and windows, return nosings on stair treads. You could even use biscuit joinery where two closet shelves meet at a corner and avoid having to screw a cleat to the underside of one shelf to support the other (drawing facing page).

Do be careful, though, if you decide to buy a biscuit joiner. Beware the "Law of the Instrument." This is a theory in psychology that states: if you give a small boy a hammer, everything he encounters will need hammering. There may be some things that really don't need to be joined with biscuits. □

Kevin Ireton is editor of Fine Homebuilding. *Photos by the author.*

Biscuit-joiner manufacturers

Looking for a faster and more accurate alternative to doweling, a Swiss cabinetmaker named Henry Steiner developed plate joinery (also called biscuit joinery) back in the 1950s. Steiner founded a company to manufacture slot-cutting machines, called plate joiners, and the plates (or biscuits) to go with them. The company is called Steiner Lamello, Ltd. ("lamello" comes from the German word *lamelle*, meaning "thin plate"), and despite the fact that nine other companies now make plate joiners, the Lamello machine is still considered by many woodworkers to be the "Cadillac" of plate joiners. The following list includes all the companies that make plate joiners. Prices range from under $150 to over $600. So if you decide to buy a plate joiner, be prepared to shop around. —K. I.

Delta 32-100 (bench-top model)
Delta International Machinery Corp.
246 Alpha Dr.
Pittsburgh, Pa. 15238
(800) 438-2486

Elu Joiner/Spliner 3380
Black & Decker, Inc.
P. O. Box 798
Hunt Valley, Md. 21030
(800) 762-6672

Freud JS 100
Freud, Inc.
P. O. Box 7187
High Point, N. C. 27264
(919) 434-3171

Kaiser Mini 3-D
(dist. by W. S. Jenks & Son)
1933 Montana Ave. NE
Washington, D. C. 20002
(202) 529-6020

Lamello (3 models)
(dist. by Colonial Saw Co., Inc.)
P. O. Box A
Kingston, Mass. 02364
(617) 585-4364

Porter-Cable 555
Porter-Cable Corp.
P. O. Box 2468
Jackson, Tenn. 38302
(901) 668-8600

Ryobi JM100K
Ryobi America Corp.
5201 Pearman Dairy Rd.
Suite 1
Anderson, S. C. 29625
(800) 226-6511

Skil 1605:02
Skil Corp.
(subsidiary of
Emerson Electric Co.)
4300 West Peterson Ave.
Chicago, Il. 60646
(312) 286-7330

Virutex O-81
Rudolf Bass Inc.
45 Halladay St.
Jersey City, N. J. 07304
(201) 433-3800

Survey of Finish Nailers

The new trim nailers are lighter, smaller and more powerful than ever

by Jim Britton

When I broke into the trades in 1973, I could count the different brands of pneumatic trim nailers on one hand. Back then, plenty of professional carpenters hung doors, trimmed windows and affixed baseboards with a claw hammer and a pouch full of 8d finish nails. Not anymore. Today, a contractor simply cannot compete without a decent complement of air nailers. That's because a pneumatic nailer not only speeds the work, it also makes for a better job. To tell the truth, I don't even keep hand-driven nails on my truck anymore.

To fill the demand for trim nailers, tool companies new and old, domestic and import, have offered up more than two dozen finish nailers for the pneumatically enhanced carpenter to choose from. And the tools aren't just for pros. The weekend builder/woodworker can also take advantage of the increased quality and convenience of a finish nailer without necessarily forking over the $400 or so that the most expensive nailers fetch. Imported tools that sell for half of that amount can let anybody join the ranks of the pros and leave the hammer tracks behind.

In this article I'll talk about the pros and cons of the details that are a part of every finish nailer. And I'll also give you my impressions of the 23 nailers I evaluated in the course of my work on various job sites earlier this year.

First, choose between 15-ga. or 16-ga. nails—Pneumatically driven finish nails come in two sizes: 15 ga. and 16 ga. Both kinds are available in lengths ranging from ¾ in. to 2¾ in. The 15-ga. nails are slightly larger in dia.—about 0.069 in. to 0.072 in. The 16-ga. nails are about 0.057 in. to 0.060 in. Which should you choose? I use my 16-ga. nailer for installing the thinner trim materials, such as ⅜-in. thick baseboards. I'll also use the 16-ga. nailer for affixing ¾-in. trim, but only when the nails aren't asked to span a gap or a soft material, such as drywall, in order to get to the substrate. For example, if I'm hanging a door and the nails have to span a gap between the jamb and the trimmer, I'll use the 15-ga. nailer. The extra beef in the larger nail is better-suited to the cantilever action of spanning a gap, especially when the nails have the variable loads imposed by a door on them.

The thinner the nail, the less likely it is to split wood. Therefore, it makes sense to use the 16-ga. nailer when running casings or window aprons. But there's a catch: Thinner nails are more likely to be deflected off course by the grain pattern in the wood. This result usually happens with chisel-point nails when the cutting edge of the nail is parallel with the grain. Solution: Rotate the nailer 90° so that the cutting edge is perpendicular to the grain.

In most cases, 16-ga. nailers have straight magazines (bottom photo), and 15-ga. nailers have angled magazines (top photo). The angled magazine allows the generally larger 15-ga. nailer to behave more like a smaller tool, poking into corners where it otherwise might not fit.

Two basic types of magazines—The strips of nails are held in the tool's magazine, the long, thin compartment below the handle. Magazines are either closed or open, rear load or top load. Most of the variations take place in the 16-ga. nailers because the rectangular strips of nails are easy to load any which way. The angled, 15-ga. strips of nails, on the other hand, are al-

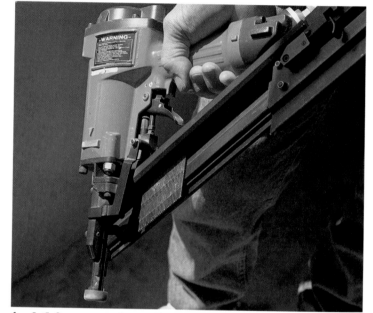

Angled for compactness. Because of their angled magazines, 15-ga. nailers fit into the same confined spaces as smaller 16-ga. nailers. This Fasco BA-65 loads from the side; a magnetic strip helps to retain the nails.

16-ga. nailers are the smallest in this class. Because they fire a smaller nail than their 15-ga. brothers, 16-ga. nailers can be packed into a smaller package. The Paslode 3250 F16 (above) has a top-load magazine.

most always fed into the magazine from the rear. To load nails into a tool with a closed magazine, you pop the latch on the magazine's cover, pull back on the magazine's spring, load the nails in from the side and close the cover. The magazine's spring is now ready to advance the nails into the firing position. This system works well if the magazine has a strip magnet built into it to retain the nails as they are loaded (top photo, facing page). But without the magnet, nails can sometimes bunch up and fall out as the spring is returned to position. My favorite magazine style is the top-load design (bottom photo, facing page). This type allows for quick loading because there are fewer parts to fiddle with, and the nails are held in their track as the spring is returned to position.

Some contact points are better than others—A pneumatic nailer engages the work by way of the contact point at the business end of the tool. As you push down on the nailer, the contact point disengages the tool's safety mechanism so that pulling the trigger fires the nail.

When I'm running trim that has a profile to it, I prefer a contact point like the one on the 16-ga. Airy ATF 0350 (photo 1, right). This rounded, horseshoe-shaped loop tucks into milled details well, allowing the nail to be driven all of the way below the surface of the trim. When I do production work with this type of contact, I slide the nose of the nailer along the trim, indexing the contact against the milled edges. The method works because the rounded metal loop rarely mars the wood.

Cushioned contacts (photo 2, right) also reduce marring, but on profiled stock a cushioned contact may make it difficult to drive the nails all of the way. Cushioned contacts are best used for installing flat stock, such as baseboards, where you want to be able to press the workpiece tight to the drywall. Remember, all cushions are removable, and this feature may allow for better nailing of detailed moldings. My advice is to avoid the bent sheet-metal contacts that some manufacturers favor (photo 3, right). The prongs on this type of contact gouge and ding all but the hardest of woods.

Consider the firing sequence—The nailer's contact is part of a safety mechanism that prevents accidental firing. In order to fire a nail, a sequence of steps has to be followed. The most common firing sequence is the contact trip. This arrangement allows the tool to fire anytime that the trigger is pulled and the contact piece is depressed. So if you keep your finger on the trigger and simply push the tool down, it will fire. When done repeatedly without releasing the trigger, this process is called bounce nailing, and it is often used by framing crews to install sheathing.

My view is that bounce nailing is of dubious value, especially for trim work where accurate nail placement is critical. Most nail jams occur during bounce nailing. It should be noted that contact-trip nailers don't have to be bounce nailed. They will drive a single nail. Another firing sequence is called the sequential trip. With this method, the contact must be depressed first, and then the trigger can be pulled to drive a single nail. The trigger must then be released before another nail can be fired. Of the tools surveyed here, most have contract-trip sequences. Only the Stanley-Bostitch N60FN and the Duo-Fast LFN-764 use the sequential-trip firing sequence.

Nailer-survey ground rules

As an interior-trim contractor, my job site is my laboratory. The tools evaluated in this article were used on a variety of projects. Some were fancy stain-grade installations, and others were paint grade, with plenty of medium-density fiberboard (MDF) trim.

For the power test, I drove nails through ¾-in. MDF into 2x Douglas fir. To see if the tools left dents in trim, I used them on pine casings. My air source was a twin-tank professional compressor rated at 3.5 CFM @ 125 psi, with no more than 75 ft. of hose. With the exception of the two oiless nailers, I put two drops of 20-weight nondetergent oil into the air intake of each nailer prior to use (never use detergent oil in a nailer; it will degrade the O-rings).

The nailers were tested in a temperature range of 20°F to 75°F, with no temperature-related failures. However, I do recommend warming nailers before winter use. Most makers will advise a thinner oil in colder weather. Also, it is important to note that I'm a middle-aged man of medium build, with average strength and medium-size hands. You should make your judgments accordingly.

Tool survey begins on the next page.

The job determines the contact type. On profiled trim, a rounded loop contact such as that on the Airy ATF 0350 (photo 1) slides along the work without denting the trim. A cushioned contact (photo 2) is best used on flat stock that has to be pressed against the substrate before nailing. The sharp prongs on a bent sheet-metal contact (photo 3) will ding soft trim, such as pine. A flip-open nosepiece (photo 4) makes for easy jam removal.

Adjustable exhaust ports are a convenience. If you have to work in a dusty area, you'll appreciate the exhaust port on this Stanley-Bostitch N60FN. It can be directed with a quick twist.

Cordless nailing for pickup jobs. Paslode's Impulse nailer is powered by an internal-combustion engine, so it needs no compressor. It's ideal for quick jobs that don't warrant schlepping compressors and air hoses.

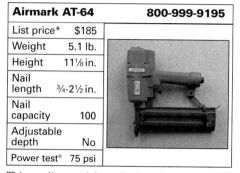

Airmark AT-64	800-999-9195
List price*	$185
Weight	5.1 lb.
Height	11⅛ in.
Nail length	¾-2½ in.
Nail capacity	100
Adjustable depth	No
Power test°	75 psi

This medium-weight nailer has decent balance and a comfortable grip. Its closed side-load magazine is simple and sturdy, but it lacks a magnet for retaining nails during loading. At 75 psi this nailer clobbered 2-in. nails through ¾-in. MDF and into fir. The major drawback to this import is the bent sheet-steel contact piece. It marred pine and tended to dig into the soft wood if the tool was slid along the trim to the next nailing point.

Airy ATF-0350	800-999-9195
Woodtek 864-374	800-645-9292
List price*	$340
Weight	4 lb.
Height	9⅞ in.
Nail length	¾-2 in.
Nail capacity	100
Adjustable depth	No
Power test°	100 psi

This nailer is compact and light, and has the uncanny ability to drive both 16-ga. and 15-ga. nails. Used exclusively on a stain-grade job, I found this tool to have the best noncushioned contact piece of any nailer tested (photo 3, top right, p. 25). I could slide this nailer down the wood without losing contact and have a perfectly placed nail almost every time. On the other hand, I did not care for the plastic-composite magazine. Static electricity seems to attract dust to the magazine, where it then sticks.

This tool has a single-screw nose for jam removal, which is slower than other flip-open models. And the tool comes with a spare piston-driver assembly, a handy part to have on hand. I wish other manufacturers included extra drivers in their packages.

Atro Monza 64	800-284-5347
List price*	$400
Weight	4.5 lb.
Height	10½ in.
Nail length	1¼-2½ in.
Nail capacity	100
Adjustable depth	Yes
Power test°	100 psi

Along with Senco, Atro is the only other company to offer oilless technology. This Italian nailer is billed as a 16-ga./15-ga. tool, but I couldn't get it to behave as such. The open rear-load magazine works with 16-ga. nails but not with the fatter 15-ga. nails. I couldn't find any 15-ga. Atro nails, so I loaded up the tool with some Stanley-Bostitch nails, which had the same 25° angle. But the Stanley-Bostitch nails wouldn't advance properly. Evidently, their heads are too big. A call to the Atro folks confirmed this. They told me that only 15-ga. Atro nails would work in the tool. Unfortunately, Atro doesn't make any 15-ga. nails at this writing. In light of this Catch-22 circumstance, consider this to be a 16-ga. nailer.

The Monza comes with a large, cushioned-loop contact. It's so large that it obscures the work and keeps the nailer from getting close to detailed moldings. Atro does offer an optional contact. It provides better visibility, but it increases marring. The depth adjustment is a hassle requiring wrenches. But the nosepiece has a single clip for quick jam removal.

Craftsman	Sears stores
List price*	$200
Weight	5.4 lb.
Height	10 in.
Nail length	¾-2 in.
Nail capacity	100
Adjustable depth	No
Power test°	90 psi

Sears' nailer uses a closed side-load magazine that required careful alignment of the nails during loading. But once the nailer was loaded, I was able to zoom on trim. The cushioned contact prevented marring, but it obscured the view a little bit. The contact wants a daily drop of oil on the point where it travels through the nose to the trigger. At $200, this Taiwanese import is a good value.

Duo-Fast LFN 764	800-752-5207
List price*	$340
Weight	5.1 lb.
Height	9 in.
Nail length	¾-2 in.
Nail capacity	120
Adjustable depth	No
Power test°	70 psi

The LFN is small and feels a little heavy. But its compact size and comfortable suedelike grip make this nailer easy to maneuver. The closed magazine with magnetic retention is as nice as they come. My only beef with this nailer is that the nails exit the tool well behind the contact piece. I got familiar with this quickly enough but still had some trouble nailing casings adjacent to hinges. The Duo-Fast brand nails come in strips of 60, for a total load of 120. This tool is solid. It'll outlast your pickup.

Fasco FN70	800-239-8665
List price*	$394
Weight	5.5 lb.
Height	11⅛ in.
Nail length	¾-2¾ in.
Nail capacity	100
Adjustable depth	No
Power test°	70 psi

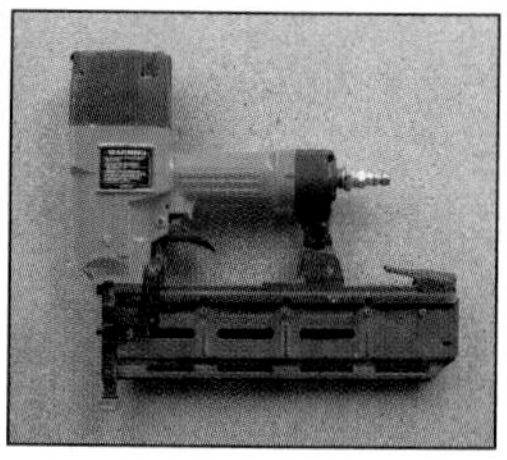

This Italian nailer is a heavyweight among the 16-ga. tools, but terrific balance offsets its weight. It has remarkable power. I used it on one of my stain-grade jobs and had it countersinking 2-in. nails into solid pine at a surprisingly low 65 psi. The Fasco performed similarly in the MDF power test. My only gripe is with the too-stiff contact spring, which complies with tougher European safety standards. With the tool overhead, I had difficulty depressing it every time. This nailer uses a semiclosed magazine with magnetic retention. Its sturdy construction and careful machining suggest this tool will last for many years.

*Retail and mail-order outlet prices are often far less than list price.
°The minimum line pressure required to countersink a 2-in. nail through ¾-in. MDF into Douglas fir.

Grizzly G2413	800-541-5537
List price*	$225
Weight	5.9 lb.
Height	10 in.
Nail length	1¼-2½ in.
Nail capacity	100
Adjustable depth	No
Power test°	110 psi

This chunky 16-ga. nailer is as heavy as it looks. Compared with other tools in its price range, the Grizzly is crude. When I first hooked it up, it would not set a 1¾-in. nail through ⁹⁄₁₆-in. MDF, through drywall and into a fir stud. I filed the nose some, which simulated lengthening the driver, but still the nailer had to be held perfectly square to the work to drive and set the nail. The Grizzly's slippery enamel finish and poor balance make it hard to hold. And the bent sheet-steel contact mars work and digs in. There are plenty of better nailers in this price range.

Haubold SKN 64 A/16	800-437-9818
Kihlberg SKN 64 A/16	800-437-9818
Hilti FBN212A	800-879-8000
List price*	$575
Weight	4.9 lb.
Height	9⅝ in.
Nail length	1³⁄₁₆-2½ in.
Nail capacity	105
Adjustable depth	No
Power test°	110 psi

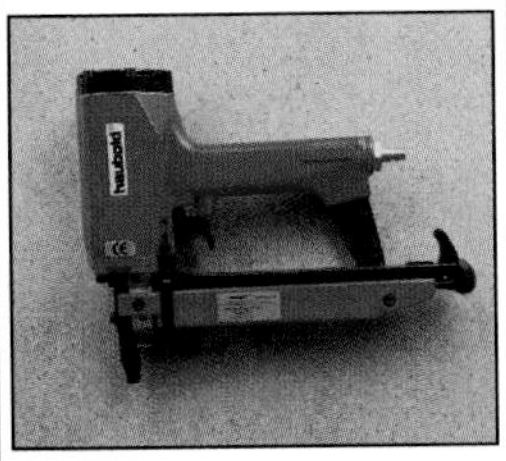

This beautifully cast nailer is a gorgeous example of German workmanship. However, it suffered the same malady as the other tools using a bent sheet-steel contact. Specifically, the tool did not set nails when the nailer was slightly canted. The contact cushion compounded this problem. By removing the cushion, nail penetration was better, but the contact piece marred the pine. This nailer has an open, top-loading magazine. It is nicely machined, and the nails slide smoothly inside it. But ergonomically, the magazine is too close to the handle. It's hard to hold the tool by its handle while loading it.

The power test revealed another weakness but more likely confirms my suspicion that the contact prevents proper nail driving. I had to set the line pressure to 110 psi to drive 2-in. nails into the test medium. With the cushion removed, I was able to reduce the line pressure. The tool is medium weight and well-balanced. It feels great in the hand all day long. Shoots all brands of 16-ga. nails.

Hitachi NT65A	800-706-7337
List price*	$749
Weight	4.4 lb.
Height	10⅛ in.
Nail length	1-2½ in.
Nail capacity	150
Adjustable depth	Yes
Power test°	90 psi

This handsome, light, well-balanced tool has a good deal of power. I drove nails at as little as 80 psi, and the contact tip is gentle on the work. The top-load open magazine holds a bonus supply of 150 nails, but my test nailer had a major problem. The nail pusher jammed nearly every time that the tool got down to about 50 nails. It would work very well for 100 or so nails. Then the pusher would slip off of the nail strip and jam. It appears that the spacers that set the width of the nail slot are too thick. With extra space in the magazine, the pusher can leave the nails. According to Hitachi, they've taken steps to correct the situation, and they will fix older nailers at no charge.

Jet JDPN-671.4	800-274-6848
List price*	$331
Weight	5.7 lb.
Height	9¾ in.
Nail length	1-2½ in.
Nail capacity	110
Adjustable depth	No
Power test°	100 psi

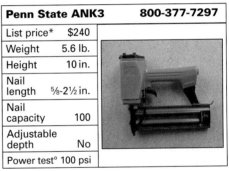

This tool is a counterfeit of the Haubold, and therefore it suffers from the same maladies as the much prettier German nailer. The bent sheet-steel contact mars and digs into wood. The handle is too close to the magazine, making loading inconvenient. Although it is a copy of a German tool, the casting and machining are not even close. Like the Haubold, the tool is compact and a bit heavy but well-balanced.

Paslode 3250 F16	800-323-1303
List price*	$360
Weight	4.5 lb.
Height	10⅛ in.
Nail length	¾-2½ in.
Nail capacity	150
Adjustable depth	Yes
Power test°	95 psi

Light and well-balanced, this nailer feels great in your hand. It, too, has a 150-nail, top-load magazine. Paslode has used this design for years, dating back to the venerable Mustang nailer. The Paslodes like a higher line pressure, and I found 100 psi to be the reliable minimum.

A clever quick-release nosepiece allows the nose to be opened in a heartbeat to clear a jam. The only thing that I didn't like about this nailer was the side play in the nonmarring contact piece. It wiggles a bit, making fast and precise nail location a little tricky.

Penn State ANK3	800-377-7297
List price*	$240
Weight	5.6 lb.
Height	10 in.
Nail length	⅝-2½ in.
Nail capacity	100
Adjustable depth	No
Power test°	100 psi

To my way of thinking, the Penn State ANK3 is another Stone-Age nailer. It appears to have been chiseled out of an iron ingot. Like the Grizzly G2413, Penn State's 16-ga. nailer has trouble setting nails consistently. If you cant the nailer to the side, the nails sometimes end up protruding from the work. The ANK3 also has a sheet-metal contact tip, which digs into the work and mars the wood. This nailer is a poor value.

Tool survey continues on the next page.

Good nails, bad nails

Tool manufacturers claim that you should use only their nails. Nonsense. I had six brands of 16-ga. nails on hand for my tests, as well as generic nails from Asia, and they were all interchangeable with the 16-ga. nailers. But nails are not of equal quality. With one exception, the best nails are made in America. The only other nails that compare are the Fasco nails (or Beck nails), which come from Austria. Steer clear of nails that have voids in the strips (photo right). A void will cause a misfire, followed by a jam. Also, make sure the strips are straight. A curved strip can cause friction, hindering the nails as they advance. A jam can result.

This gap can cause a jam.

Spotnails HB1640	800-873-2239
List price*	$380
Weight	5.3 lb.
Height	10¼ in.
Nail length	1-2½ in.
Nail capacity	100
Adjustable depth	No
Power test°	80 psi

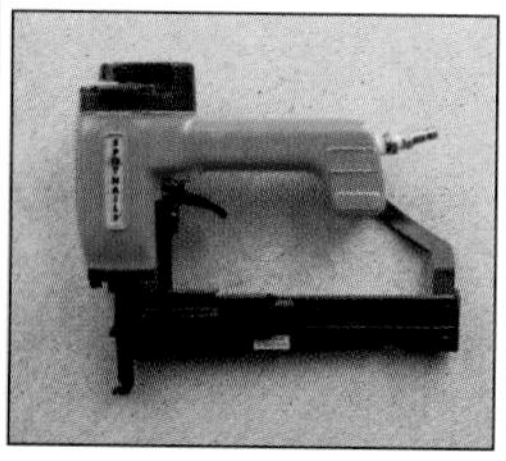

The 1640 is a curious nailer: It's both heavy and compact. The first things that I noticed when using the tool were the faraway feel of the trigger, and its long travel. At first, I would be pulling the trigger and waiting for the nail to shoot. Then I would pull a little farther, the nail would fire, and I'd flinch. After I'd gone through a box of nails, however, I got used to the tool.

This nailer has a simple, top-loading open magazine that's easy to use. I liked the exposed pusher spring because it's easy to clean. The contact piece is the common front loop, but it's made out of flat stock and doesn't slide along the work very well. Still, the tool has good power and a no-nonsense likability.

Spotnails FB1632	800-873-2239
List price*	$268
Weight	3.4 lb.
Height	8½ in.
Nail length	¾-2 in.
Nail capacity	100
Adjustable depth	No
Power test°	85 psi

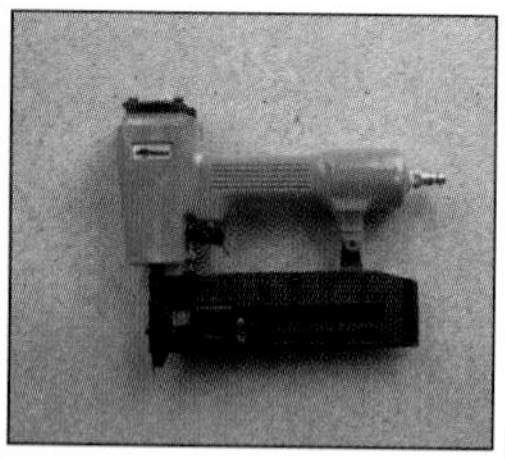

If you like compact, then you'll love the 1632. This nailer is by far the smallest one in the survey. It's about the size of most 18-ga. brad nailers. But it still has incredible power. Pulling the nail pusher back and over the loaded nails is awkward at first, but I found it easier to reload after I'd run about 1,000 nails through the tool. The contact piece is too small, and as a consequence, it mars the pine trim. Fortunately, the tool has a soft contact spring.

Airy ATH-0565T	800-999-9195
Penn State ANK5	800-377-7297
Woodtek 832-392	800-645-9292

List price*	$398
Weight	5.1 lb.
Height	10¾ in.
Nail length	1-2½ in.
Nail capacity	100
Adjustable depth	No
Power test°	110 psi

This import from Taiwan is a surprisingly good tool. Its medium weight, excellent balance and smallish head make the nailer maneuverable. It has an excellent, nonmarring, cushioned contact piece that provides good visibility. The open, rear-load magazine was my favorite of all of the angled nailers tested. The magazine is simple and loads easily. The nails almost glide into their slot. The nose opens with a clip for easy jam removal. My only complaint about this nailer was that its tiny screws would loosen and fall out. This problem could be solved by using a thread-locking fluid prior to operation.

DeVilbiss AFN 5	800-888-2468
List price*	$280
Weight	5.9 lb.
Height	11 in.
Nail length	1-2½ in.
Nail capacity	100
Adjustable depth	No
Power test°	110 psi

The U. S.'s largest maker of air compressors now imports the AFN 5. It is a Senco knockoff, and a pretty good one. It is medium-heavy, but well-balanced. The mar-free cushion tip tended to fall off the contact, which made the nailer jam frequently. I solved the problem by gluing the tip back on with a dab of construction adhesive. If the tool does jam, it can be easily cleared by flipping open the front of the nosepiece. The open, rear-load magazine is made of steel, which is reminiscent of the older Senco SN1. This nailer loads easily and has reasonable power. For hard materials, though, you have to increase the line pressure to the maximum.

Duo-Fast HFN 880	800-752-5207
List price*	$400
Weight	7.7 lb.
Height	11¾ in.
Nail length	1½-2½ in.
Nail capacity	100
Adjustable depth	No
Power test°	65 psi

Holy smoke! This nailer easily won the most-powerful honors. It nailed the power test at 65 psi, with a 2½-in. nail through a double layer of MDF into a fir stud. The straight, closed magazine includes magnetic nail retention. It opens conveniently with a nicely located front-release latch. At a whopping 7.7 lb., it is also the heaviest tool of the bunch. But its excellent balance and a suede-like grip make it easy to hold.

As with the Duo-Fast LFN, this tool has the nail exit the nose behind the contact piece, which makes for some awkward nail placement. The cushioned contact, with a built-in hard-rubber pivot, does mar some. The pivot allows the contact to be square to the material, even when the nailer isn't square to the material. Because of its great power and solid workmanship, I think this tool is best-suited to a shop situation where its lack of maneuverability won't count against it.

Fasco BA-65	800-239-8665
List price*	$358
Weight	6.4 lb.
Height	11⅝ in.
Nail length	1-2½ in.
Nail capacity	100
Adjustable depth	No
Power test°	110 psi

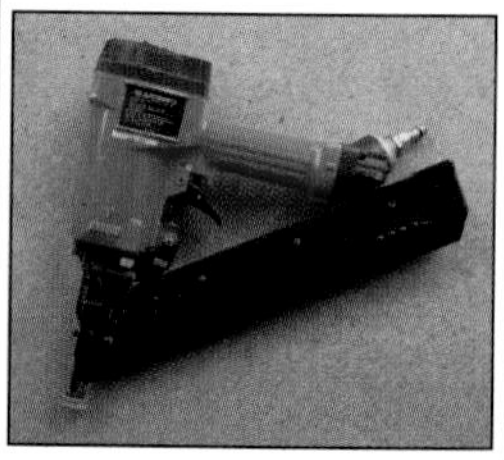

The Italian penchant for durability and design is evident throughout this nailer. It is a heavy tool, but it is well-balanced. The closed magazine includes a magnetic retention strip.

This nailer won't shoot Senco nails, which is a potential drawback because Senco nails are widely available. I got out my protractor to see why. The Fasco nails read approximately 34½°. The Senco nails read about 35½°. Interestingly, the Fasco worked fine with the All-Spec nails, which are also collated at 34½°. The BA-65 has good power, but I had to raise the line pressure to the maximum for hard materials. The air-supply port is smartly angled for easy connection. As I did with its little brother, I found the BA-65's contact spring to be too stiff.

*Retail and mail-order outlet prices are often far less than list price.
°The minimum line pressure required to countersink a 2-in. nail through ¾-in. MDF into Douglas fir.

Makita AF 631	714-522-8088
List price*	$425
Weight	5.6 lb.
Height	12⅛ in.
Nail length	1¼-2½ in.
Nail capacity	100
Adjustable depth	Yes
Power test°	90 psi

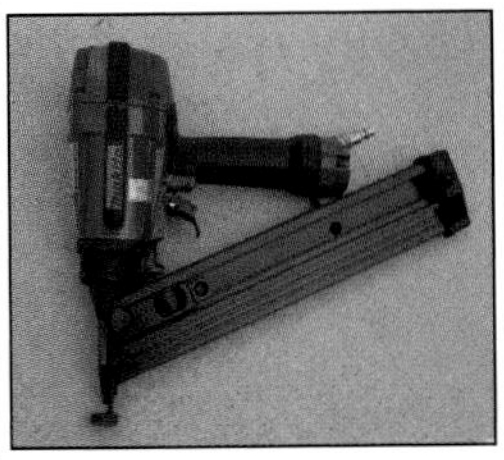

The new Makita nailer works beautifully and looks great. It is similar to the Stanley-Bostitch N60FN and uses the same 25° nail strips. Makita will have its own nails available when it introduces the tool in the fall. While a bit heavier than the Bostitch, the Makita is well-balanced. It has a comfortable handle, a nice trigger and a nine-setting depth adjustment. This nailer has the best quick-open nosepiece that I have seen. It has no screws or clips. To open, simply push the cover up and open.

A clever new feature is in the magazine. With an open, rear-load magazine, it looks pretty basic. But hidden away is an override that will not allow the tool to fire when it is empty. This feature will save on tool wear. The feature also saves unneeded puttying.

Makita has paid great attention to detail. The workmanship is as high as I have seen. I enjoyed the testing so much that I trimmed three houses with this prototype.

Senco SFN 1	800-543-4596
List price*	$514
Weight	4.4 lb.
Height	9⅝ in.
Nail length	1-2 in.
Nail capacity	104
Adjustable depth	Yes
Power test°	90 psi

Forget about squirting oil into this nailer. After introducing the first oilless nailer in 1985, Senco has the technology perfected. This tool is light and easy to maneuver, with great balance. The plastic-composite magazine easily loads from the rear, but dust tends to cling to it. This nailer now includes the excellent depth-adjustment dial from the newer SFN 40. The SFN 1 has good power and is quiet. The quick-open front and the cushioned contact with "cross hairs" make this tool one of the best.

Senco SFN 40	800-543-4596
List price*	$580
Weight	4.9 lb.
Height	11⅜ in.
Nail length	1¼-2½ in.
Nail capacity	104
Adjustable depth	Yes
Power test°	80 psi

The SFN 40 replaces the cumbersome SFN 2. This new nailer has all of the features of the SFN 1 plus greater nail-length capability at a cost of only an additional half-pound in weight. The depth adjustment is a brilliant self-locking cam. The SFN 40 has a quick-change exhaust port for directing the exhaust. For all of the performance, this tool is quiet and has almost no recoil. It's among the top tools.

Stanley-Bostitch N60FN	800-556-6696
Stan-Tech SDN 15 BR	**800-343-1234**
List price*	$550
Weight	4.4 lb.
Height	11⅜ in.
Nail length	1¼-2½ in.
Nail capacity	100
Adjustable depth	Yes
Power test°	100 psi

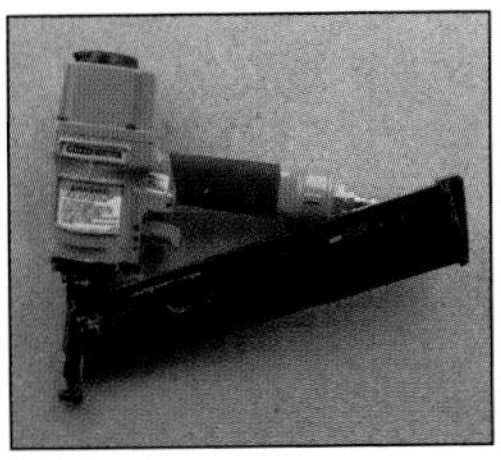

This tool is well-made; however, it seems to lack punch. To make it reliably pass the power test, I had to take the line pressure beyond the tool rating. This problem could be related to the tool's light weight. It is the lightest nailer to handle a 2½-in. nail. Beyond this glitch, I found the N60 to be beautiful, light and well-balanced. The tool's exhaust port can be easily redirected (center photo, p. 25) for those times when you're working along a floor that's covered with drywall dust.

This tool, along with the Atro and the new Makita nailer, is the only 15-ga. nailer that used nails collated at 25°. Also, these nails are a tad fragile. They tend to break apart in a tool pouch.

The Stan-Tech is the same tool, but it is painted bright blue. It's available through contractor-supply outlets.

Paslode Impulse	800-323-1303
List price*	$625
Weight	6 lb.
Height	12½ in.
Nail length	¾-2½ in.
Nail capacity	100
Adjustable depth	Yes
Power test°	n/a

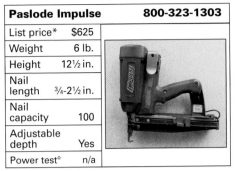

The Impulse IM250F, the only cordless nailer, is in a league of its own. This 16-ga. nailer uses a liquid hydrocarbon fuel to drive the engine that propels the nail. Although it looks bulky, the nailer weighs only 6 lb. with fuel, battery and nails.

As this tool evolves, it is sure to get smaller. But in its present configuration, it does not get into corners very well. This tool's niche is the quick job, the callback or a short punch list—any time that packing the compressor out, running some hose and running the cord would get to be too much of a hassle (bottom photo, p. 25).

Choosing a nailer

Selecting the right nailer depends on the type of work you plan to do with it and how often it will be used. Will you use it occasionally in a home shop, or every day as a professional? If you plan on part-time use, my suggestion is to get one of the imported tools. They cost about half of top-of-the-line domestic models, and they will give years of service to the part-time user.

Choose a 15-ga. nailer if you plan to nail mostly hardwoods. A 16-ga. nailer is fine for softwoods. If you've got a cabinet shop, stick with the 15-ga. nailer. Larger nails are less likely to deflect and ruin a cabinet.

For the professional trim carpenter, reliability is the most important issue. Get a top-flight nailer that has a local supplier for both parts and nails (some manufacturers even offer job-site service). If you're going to get only one finish nailer, make it a 15-ga. model. And no matter what the application, always use the best nails you can find. □

Jim Britton is a trim carpenter and contractor living in Fairfield, California. Photos by Charles Miller.

Installing Baseboard

There's a tad more to it than coping the joints

by Bob Syvanen

Baseboard installation is often done badly. Why? Probably because it comes at the end of the job, after the crowns and casings, and carpenters are anxious to wind things up so they can get on to new projects. Or it may be because it's uncomfortable work done on hands and knees, with a lot of getting up and kneeling down. But maybe it's just because a lot of carpenters don't know how to do it right.

Shapes and styles—Baseboards are used to cover any gaps that occur at the juncture of walls and floors, and they also protect the lower wall from dings and scrapes. Visually they give weight, definition and presence to the wall, working with the crown molding and corners to "frame the wall." Baseboards are usually made of the same wood that's used for trim elsewhere in the house, and they can be either hardwood or softwood. The central part of the back face of baseboard stock is partially relieved, or plowed away, like casings; this helps it lie better against the wall. Baseboards, however, are usually thinner than casing stock. This is because casing is frequently made with a rounded outside edge, and a somewhat thinner baseboard can be butted against this edge without its looking awkward.

Standard baseboard comes in a variety of shapes and sizes, and custom shapes can be made in the shop with a table saw, router or shaper. Another way of getting a unique baseboard profile is to assemble it from combinations of standard moldings, as shown in the drawing below. There's really no end to the shapes that can be achieved when two or three-piece combination baseboards are used.

Coping the joint—Many of the techniques for cutting, fitting, nailing and finishing baseboard are similar to those required for casing. But the miter joint used so frequently for casing tends to open up on the inside corners of baseboards. A much better baseboard joint for inside corners is the coped joint.

A coped joint requires a different cut on each of the two boards to be joined. It involves some miter cutting, so a backsaw and miter box or a power miter box are required. Though I've cut miters by hand for years, I like the power miter box because it's fast, but doesn't sacrifice quality. You'll also need a coping saw. This small tool with a spring-steel frame looks like a C-clamp with a wood handle, and has a slender, fine-tooth blade stretched across the mouth of the C.

Cutting the first board in a coped joint is easy—just cut it square so it fits tight into the corner of the wall. The second board is coped. To begin a coped joint, miter the board vertically, as if you were going to make an inside mitered corner. When you're done, look closely at the front edge of the cut—it will reveal the baseboard's profile, and will serve as a guideline for making the second cut on the board. I rub the edge of the cut with the edge of a pencil lead to make it more visible.

To complete the cope, support the board, front face up, so that the end to be cut hangs just beyond some solid support—a workbench, sawhorse or cricket (a cricket is a portable step turned mini-workbench). Then, with the sawblade nearly perpendicular to the bottom edge of the board, cut along the pencil line, following whatever curve is indicated (drawing, facing page, left). While cutting, incline the saw slightly to put an angle on the cut. The angle should slope away from the front surface of the board, and will help the lead edge to make good contact with the square-cut board when the two are brought together. If the baseboard has a flat top edge (like the one in the drawing), this edge should be square cut—an angle would show as a gap. If you've made the cut correctly, the end of the coped baseboard will slip right over the square-cut end of the one you installed earlier. This technique will work on just about any baseboard, and can also be used to fit ceiling molding. It sometimes requires a little adjustment with a sharp chisel or utility knife.

Installing standard baseboard—If there's a simple choice between a long, unbroken wall and a short one, I start the installation with the long one. It's easier to get a good fit with a long piece of baseboard than with a short one, and you'll see why in a moment. I also try to minimize any possibility that people will see a poorly fitted corner joint (if one happens to slip into the job). To do this, I like to install the first length of baseboard on the side of a room that's opposite the door. The baseboard on the adjacent wall will conceal the imperfect joint so that it won't be visible when someone first enters.

Let's assume that you've chosen to start on an unbroken wall that can be fitted with a single length of baseboard. Begin by measuring the wall, making sure to take your measurements at the floor level. Walls aren't usually in perfect plumb, so the measurement will vary depending

When baseboard must be fit to door casing, a measuring block (photo left) makes it easy to mark the baseboard to length. Built from scrap wood to fit the particular baseboard being installed, it is placed against the outside edge of the casing and over the baseboard, and a cut line is then scribed on the baseboard.

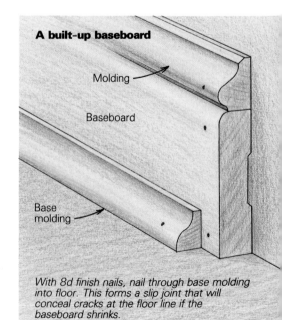

A built-up baseboard

Molding

Baseboard

Base molding

With 8d finish nails, nail through base molding into floor. This forms a slip joint that will conceal cracks at the floor line if the baseboard shrinks.

Photo: Pat Syvanen

on where you take it. Select a straight length of baseboard and clean up one end by cutting it square. Use the wall measure you took to find the other end and cut it square also. The cut should allow a snug fit, particularly at its top edge where it will be visible. Test the fit on the wall, trim off a tad if necessary and nail the piece in place. If you have to splice lengths of baseboard on long walls, a vertical scarf joint (overlapping 45° miters) is the best to use.

Nail through the baseboard and into the studs, using 8d finish nails top and bottom. Use as many nails as necessary to get the baseboard to pull tight to the wall. Studs can be located before setting the base in place by probing with a nail driven through the finished wall (the baseboard will cover the holes). Mark the locations above where the top edge of the baseboard will be with a light pencil mark so they won't be obscured when you set the baseboard in place.

With this done, you can begin work on the adjacent length of baseboard. This one gets a coped cut at one end and a square cut at the other (this same combination of cuts is used when a baseboard has to be fit between a door casing and another baseboard). First, cope the end that will butt against the board you just nailed in place, and check it for a good fit. Then measure to locate the square cut at the other end. The fit should be snug, but not so tight that it cracks the plaster or drywall when you nail it in place.

To get just the right amount of snugness on long boards, cut them a tad long so that the middle of the board will be bowed out from the wall when you put it in place. The amount of bow depends on the length of the baseboard you're installing, but it's usually about a finger's width, measured midway between the ends. But you shouldn't have to force it—a gentle push and the baseboard should snap into place as it nears the wall. Take particular care when you're fitting baseboard to the casing of a door. As you spring the baseboard into place you don't want to push the casing out of position. If the fit looks good at both ends, nail it in place. Keep work-ing your way around the room until all the baseboard is in place. There may be times when you have to fit a baseboard between baseboards on opposite walls. If you do, just cope both ends and snap it into place.

For fitting baseboard to door casing, I use an L-shaped block (photo facing page) to help me mark an accurate cutline. By holding the top of the block against the edge of the casing and marking down the leg onto the baseboard, I get a precise measurement. With one end coped, slip the baseboard into place, allowing it to run slightly long past the casing on the other end. Place the block against the outside edge of the casing and over the baseboard, and mark a vertical line for the square cut. A cut made a tad outside this line will give you just enough extra length to spring the board into place.

Other joints—Coping doesn't work on all baseboard profiles. Simple rectangular baseboard, the kind with a rounded upper corner, just doesn't look good with coped joints because the abrupt curve makes for a very fragile overlap on the adjacent baseboard. Instead, I use a combination butt and miter joint for inside corners (drawing, below right).

The trick here is to make a mitered lap joint at the top edge of butting baseboards. Begin with a square cut at the end of one baseboard. Miter the top edge with a finish saw, cutting to the point where the rounded corner ends, and follow this with a cut at 90° to the first. This will release a triangular piece of wood. At the end of the second baseboard, make a miter cut that corresponds to the first, again just to the bottom of the rounded edge. Then turn the baseboard over. Working from the bottom edge, make a square cut that's angled just a tad away from the front of the workpiece. This cut should be stopped at a line scribed off the baseboard already in place, because there isn't much room for adjusting this joint. This is a much better-looking joint than a butt joint or a coped joint, though a butt joint can be okay when molding is used on top of it.

Outside corners—Outside corners on baseboard are always mitered. Since walls rarely make a perfect corner, I always make trial cuts to find the right angle for the miter. For a good fit, the angle cut on both pieces should be the same. I make my first cut long and gradually trim it to perfection using the miter box or a block plane. (With a miter box, shim cardboard or even plane shavings between the miter-box fence and the back of the baseboard; make fine adjustments to the saw angle by moving the shim away from or closer to the saw blade.) Outside corners are cross-nailed to lock the joint in place.

More tips on nailing—Occasionally you will be faced with a situation where the baseboard has to be pulled in against the wall, but there isn't any stud to nail into behind it. As an alternative, drive a 16d finish nail through the baseboard, angling it down and out so that it catches the 2x4 bottom plate in the wall. When the nail is set, the baseboard will be pulled snug against the wall.

In similar situations at inside corners and door openings, 16d finish nails again come to the rescue. Just angle the nail until you hit something solid. Keep it a few inches from the end of the baseboard and predrill the hole before you nail to prevent splitting. If you are framing up a new house, it's a good idea to install short lengths of 2x4 baseboard blocking (offcuts or scrap pieces) at all inside corners and at each side of door openings.

If you're installing a two or three-piece baseboard, the lower molding should be nailed to the floor, not to the baseboard. If it isn't, any shrinkage of the baseboard will pull the molding away from the floor and expose an unsightly crack. When it's nailed to the floor, molding serves as a slip joint, concealing any shrinkage cracks. Paint the baseboard before installing this molding so an unpainted strip won't show up after the baseboard has shrunk. □

Bob Syvanen, of Brewster, Mass., is a consulting editor of Fine Homebuilding *magazine.*

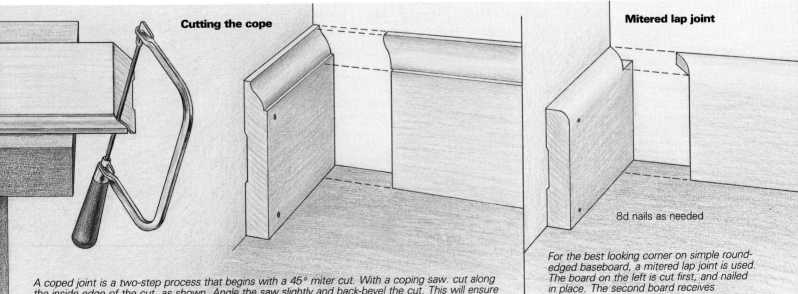

Cutting the cope

A coped joint is a two-step process that begins with a 45° miter cut. With a coping saw, cut along the inside edge of the cut, as shown. Angle the saw slightly and back-bevel the cut. This will ensure that the visible front edge of the coped baseboard will fit tightly against the adjoining baseboard.

Mitered lap joint

8d nails as needed

For the best looking corner on simple round-edged baseboard, a mitered lap joint is used. The board on the left is cut first, and nailed in place. The second board receives corresponding cuts.

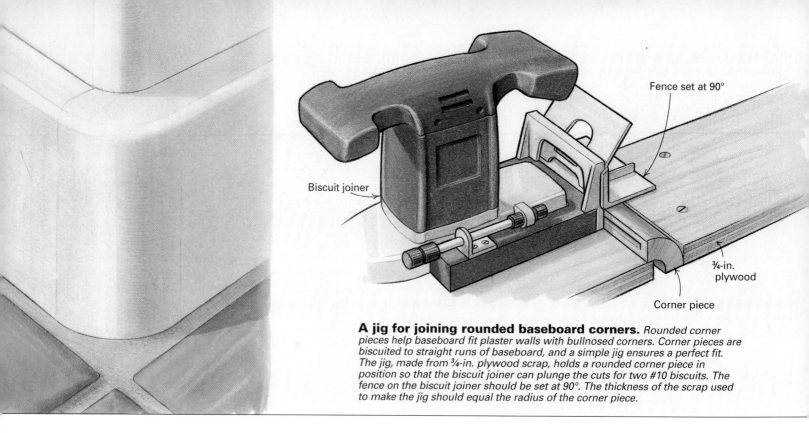

A jig for joining rounded baseboard corners. *Rounded corner pieces help baseboard fit plaster walls with bullnosed corners. Corner pieces are biscuited to straight runs of baseboard, and a simple jig ensures a perfect fit. The jig, made from ¾-in. plywood scrap, holds a rounded corner piece in position so that the biscuit joiner can plunge the cuts for two #10 biscuits. The fence on the biscuit joiner should be set at 90°. The thickness of the scrap used to make the jig should equal the radius of the corner piece.*

Labels on illustration: Biscuit joiner; Fence set at 90°; ¾-in. plywood; Corner piece

Curved Baseboard Corners

Biscuit joinery and a simple jig solve a perplexing trim problem

by Eric Blomberg

Thin-coat plaster is increasingly popular for interior walls in northern California where I work, but the variable textures and the bullnose corners of these walls can raise havoc when it's time to apply trim. I discovered this a couple of years ago when the crew I was working on started to install baseboard in a house near Santa Rosa, California. Each outside corner on these coarse-textured walls had a ¾-in. radius. Mitering the baseboard around these corners was out of the question. Square corners would look bad juxtaposed with the bullnose walls, and mitered corners would leave an awkward gap between the wall and the baseboard. Pondering paradigms led us to our current solution.

Our technique requires only a biscuit joiner, a simple jig and rounded corner stock that we have run off at a local mill shop. The back of the rounded corner pieces are concave, so they fit snugly against the radius of the plaster walls at the corner. That eliminates the triangular gap that would result if the baseboard were mitered at 45°. In this house we butted the 1x6 maple baseboard into the rounded corner pieces, so the baseboard follows the wall cleanly, even if the corner isn't exactly 90°.

It took trial and error to get the technique right. We realized that rounded corner stock was part of the answer; the trick was learning how to join the baseboard and the corner pieces together cleanly. We resolved the problem with biscuit joinery, using a jig we devised. The technique is fast and effective.

Corner pieces from a mill shop—The corner pieces were created with a single pass through a multiple-head molder, producing 10-ft. and 12-ft. long pieces that we cut to length on the job. Because the shop grinds knives for each job, it could easily make radiused corner pieces. You can also make the pieces yourself (see sidebar facing page).

For baseboard that will be painted, grain direction in the corner pieces is irrelevant. That was the case on this job. But the mill shop we use also can produce corner pieces with the grain running horizontally, just like the base. That would be helpful if the trim were going to get stained, and the grain direction had to match.

Corner pieces first—We have a three-step procedure for fitting baseboard, and we start with the corner pieces. After cutting the pieces to match the height of the baseboard (drawing above), we put them in our jig and mark the locations for two #10 biscuits (biscuits come in three sizes, ranging from #0, the smallest, to #20, the largest). Using #10 biscuits with 1x6 baseboard means we can get two biscuits at each joint, and the slots aren't deep enough to break through the face of the corner pieces. (We cut the biscuit slots in the baseboard later, after all of the pieces have been cut to length.)

The trick to cutting the slots accurately is the jig; it holds the corner piece and aligns the cutter in the biscuit joiner. The jig is simply two pieces of ¾-in. scrap plywood screwed to a base. The corner piece fits between the two scrap pieces. The jig aligns the base plate of the biscuit joiner with the inside edge of the corner piece. If the corner piece is aligned correctly in the jig, the joint between the baseboard and the corner piece will be flush. It's important to cut the biscuit slots perpendicularly to the end of each corner piece; if the slots are skewed, gaps at the joints are inevitable. After marking the corner piece for slot locations, the cuts can be made on both edges. With that done, we dry-fit the corner

Drawings: Bob La Pointe

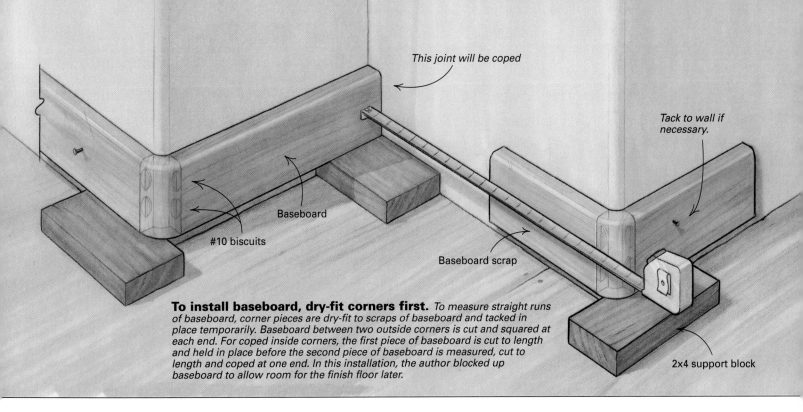

This joint will be coped

Tack to wall if necessary.

Baseboard

#10 biscuits

Baseboard scrap

2x4 support block

To install baseboard, dry-fit corners first. *To measure straight runs of baseboard, corner pieces are dry-fit to scraps of baseboard and tacked in place temporarily. Baseboard between two outside corners is cut and squared at each end. For coped inside corners, the first piece of baseboard is cut to length and held in place before the second piece of baseboard is measured, cut to length and coped at one end. In this installation, the author blocked up baseboard to allow room for the finish floor later.*

pieces to scrap pieces of baseboard and then use a router and a roundover bit on the top edges of the corner pieces so that they will match the top-edge profile of the baseboard. Once this profile has been cut, the corner pieces are ready for installation.

Measuring and installing the baseboard— The second step in the process is to set corner pieces in place temporarily and measure the straight runs of baseboard. In the Santa Rosa house, a tile floor was to be installed after the baseboard was installed, so we raised the baseboard 1½ in. off the subfloor and took our measurements for the base at this height (drawing above). To hold corner pieces in place while we measured straight runs of baseboard, we dry-fit scraps of base to corner pieces with biscuits. The assemblies could be tacked to the wall to hold corner pieces in the correct position while we took measurements.

With the straight runs of baseboard all cut, we could install all of the pieces. At each outside corner, we dry-fit both pieces of base to the corner piece to check the joints. If the fit looked good, we glued the slots, inserted the biscuits and then stuck the pieces together. If the dry fit had been perfect, we would have nailed the pieces in place right after the glue was applied.

If extra pressure were needed because one of the joints was slightly off, we let the joint set up off to one side before installing the pieces. When the glue was set, we nailed the assembly in place. It was sometimes possible, especially with shorter pieces, to assemble a three-piece corner or a five-piece U-shaped section, let the glue dry and then install it in one piece. The process may sound tedious, but it's not. Once a rhythm is established, the work flows smoothly. □

Eric Blomberg is a carpenter with Jim Murphy & Associates of Santa Rosa, Calif.

Making corner pieces on a shaper

If there's not a mill shop in your area that will make the corner pieces, you might try to make them yourself. The required tools are a table saw, a shaper and, of course, the necessary shaper knives. I wouldn't recommend a router because the bits required for the cuts would be very large.

A mill shop in my area makes all the corner stock I need. But if I were going to make my own corner stock, I'd use a three-step process on the table saw and shaper. First I'd rip the stock to the appropriate dimensions (drawing below). These cuts would establish the two faces of the corner piece that will be joined with the straight runs of baseboard.

Next, I would cut the inside curve on a shaper. The radius depends on variables like drywall or plaster thickness and the thickness of the base itself. Then I would go back to the table saw to rip the opposite (outside) corner to remove most of the waste material.

Finally, I'd use the shaper to finish up. Again, the radius required may vary. I think two passes would do it, each pass cutting half the outside radius. A little sanding will finish the job up nicely. If you try this, use material that's long enough to be machined safely. —*E. B.*

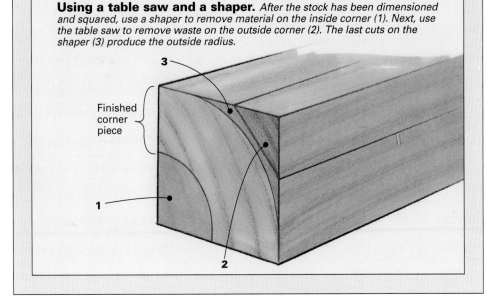

Using a table saw and a shaper. *After the stock has been dimensioned and squared, use a shaper to remove material on the inside corner (1). Next, use the table saw to remove waste on the outside corner (2). The last cuts on the shaper (3) produce the outside radius.*

Finished corner piece

3

1

2

Running Baseboard Efficiently

Simple steps help you make the most of time and materials

by Greg Smith

Let's face it. There is little or no glory in the installation of baseboard. If you want, for instance, to talk about hanging doors, you can probably find plenty of guys who are happy to grant you their expert opinions on the best tools and the most elaborate techniques. But when it comes to installing baseboard, we're back to grabbing a scrap of lumber or an unspent napkin from lunch to record measurements.

The job may go something like this: enter room, plop saw on floor, measure, cut, nail; measure, cut, nail; measure, cut, nail. And you wonder if you'll ever get to the last piece, because it seems like there is always another little piece in some nook or cranny or some space that was missed. It is a job that brings screaming protest from the knees and a hacking voice of discontent from the lungs of the person who fires the nailer that kicks up the dust from the floor adjacent to the workpiece. That may be why, when a team of carpenters is finishing a house, running baseboard is often relegated to the least-experienced person of the group or the low man on the totem pole.

The best way to deal with an unpleasant, though necessary, task is to get it done as quickly as possible. I have seen many carpenters approach the installation of baseboard in many different ways, but I had never seen a system that works very efficiently. That's why I developed a methodical approach that makes baseboard installation fast and efficient.

1. Plan your strategy—The time to run baseboard is before the painter has hidden the location of studs (assuming that you are dealing with drywall) and after the door casing, the built-ins and cabinets and the hardwood or tile floors are installed. In the areas that will be carpeted, hold the baseboard off the floor with a piece of hardwood flooring or other scrap of ¾-in. material—you won't want your beautiful work hidden by the carpet. Though I like to leave a wake of completed baseboard behind me when I am working, that's not always possible. If the bathroom floors have yet to be tiled, for example, I cut my baseboard for the room and set it aside.

2. Set up the saw—I usually set up my 10-in. power miter saw across extra-tall sawhorses (so that I don't have to bend far to see a close-up of the cut) in the biggest room on the floor level on which I'll be working. I'm not as concerned about how close I am to the area to be worked

At the wall. To run baseboard efficiently, Greg Smith measures several rooms at a time. He prepares a cut list as he goes and marks lengths and cut angles on a preprinted form.

At the saw. **With a pile of stock nearby and his list of baseboard dimensions at the ready, Smith parks himself at the saw and cuts down the list. Each piece gets a number that corresponds to the list; later on, he'll be able to mate each length with the correct wall.**

on as I am about having enough room to extend long lengths of material on either side of the saw. With my system, I don't spend a lot of time walking back and forth between my saw and the work area. If there are no large rooms, or if for some reason I cannot use one, I set up instead where I can extend stock out through a door or a window. When all of the baseboard material is spread out near the saw, I'm ready to start work.

By the way, no matter how clean the subfloor is, when you're shooting in baseboard with a nailer, the dust is going to be flying, and your mouth and nose, being close to the ground, are going to scoop up a lot of it. You might want to use a dust mask for this part of the job. In situations where I can free up my left hand, I put it under the place where air is released from the nailer. This keeps the dust from being kicked into the air.

As for what joints to cut, it's up to you whether you miter or cope. The system I use to organize the process won't change the way you work. For purposes of explanation, however, I'll use the example of mitered baseboard.

3. Install the long boards—Tackle the longest walls first—the ones that are longer than the stock you are using. These walls will require the base-

board to be spliced somewhere along its length (drawing next page, bottom). Start by cutting lengths of stock with a 45° inside miter on the left-hand side and an inside 22½° miter on the right-hand side (assuming you're working from left to right). The splice will occur at the second cut. Later on, you'll be able to take one measurement from the 22½° end to the corner of the wall to complete the wall. Go ahead and install these lengths of baseboard.

4. Take closing measurements—Now take all remaining measurements for three or four rooms at a time. Starting at the door, measure each and every length in the room. Work your way around the room in a clockwise or counterclockwise direction, whichever you prefer. The direction you choose is not as important as being consistent. Use a mechanical pencil (the kind you can get in any grocery or drug store) both for recording your measurements and for marking cutlines later on. The point will always be sharp and consistent, and you won't whittle away time carving up a carpenter's pencil with a razor knife.

Each measurement is recorded on a very simple form that I've designed and carry on a clipboard (photo facing page). The column of

blanks is sequentially numbered, and the measurements are recorded in order (drawings next page). The sequential numbers are important. They will later be recorded on the boards and used to guide you toward correct board placement. You can increase your ability to keep track of what you're doing by drawing a line between each set of measurements when you change rooms. Write the name of the room in the vertical space to the left of the column. In the example, boards #1 to #4 go in the master bedroom, boards #5 and #6 go in the closet.

The solid line to the right of the blanks represents the baseboard as you are looking at it on the wall. On the left and right ends of this line, you will write a symbol to represent the kind of cut required on each end. Use whatever symbols make sense to you—but be consistent. In the small example on the facing page, the straight, vertical line on baseboard #1 means the cut is to be a straight, square cut. This end of the board will butt against the casing around a door. The "2" on the right-hand side of the notation represents an inside 22½° cut. Baseboard #2 starts with a 22½° cut; the slash at the end of the baseboard line indicates a 45° inside cut. Baseboard #4 shows a square cut and a 45° *outside* cut (represented by the O at the end of the line). When working with a stain-grade wood, such as oak, most carpenters like to cope one end to get a tighter fit. I use a C to indicate the end to be coped. You won't find many different cuts in baseboarding, so it won't be hard to memorize the symbols you'll need. If you work with others, you might want to define your symbols on the form so that everyone will be singing from the same song sheet. As for measuring, I simply hook the tape where it's most convenient.

5. Cut each closing board—After you have compiled measurements from several rooms, take your clipboard to the saw and start cutting baseboard (photo left). Each time you cut a board, mark the backside near an end or in the middle (be consistent so that you will know where to look for it later), using the sequential number on your list. Then cross that length off your list. There is no need to write exact lengths on the board, as many carpenters do. For example, if you cut a 22½° inside cut on the left-hand side and 107⅜ in. to the right, make a 45° inside cut; turn the board over, write the number 3 and cross it off your list. Then set the board aside and begin work on board #4.

You need not cut in any particular order. One additional advantage of this method is that you can make very efficient use of your material by taking a little extra time here to avoid waste. Start with your longest pieces, then see what you can get out of the offcuts. You'll have a long list from which to choose.

6. Distribute the boards—Now that you have cut and numbered all of the pieces on your list, you can distribute the pieces. Because you have numbered them in the order in which you measured the walls, it is easy to pick up any random piece and quickly find the place where it belongs. Let's say that the first board you pick up is

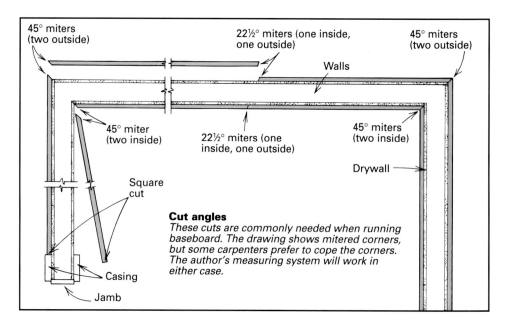

Handwritten sample (M. BED):

1. 6.5 ½ ⊢2
2. 48 5/16 2⊣
3. 60 ⅜ ⁄⊢
4. 92 ⅜ ⊢0
5. ....
6. ....

Baseboard cutlist. The sample at left shows how to use the cutlist (below). Marks at the end of each solid line indicate the cuts to be made on each baseboard.

Baseboard Cutlist

1. .... ——	1. .... ——	1. .... ——
2. .... ——	2. .... ——	2. .... ——
3. .... ——	3. .... ——	3. .... ——
4. .... ——	4. .... ——	4. .... ——
5. .... ——	5. .... ——	5. .... ——
6. .... ——	6. .... ——	6. .... ——
7. .... ——	7. .... ——	7. .... ——
8. .... ——	8. .... ——	8. .... ——
9. .... ——	9. .... ——	9. .... ——
10. .... ——	10. .... ——	10. .... ——
11. .... ——	11. .... ——	11. .... ——
12. .... ——	12. .... ——	12. .... ——
13. .... ——	13. .... ——	13. .... ——
14. .... ——	14. .... ——	14. .... ——
15. .... ——	15. .... ——	15. .... ——
16. .... ——	16. .... ——	16. .... ——
17. .... ——	17. .... ——	17. .... ——
18. .... ——	18. .... ——	18. .... ——
19. .... ——	19. .... ——	19. .... ——
20. .... ——	20. .... ——	20. .... ——

marked 18, and you see that it is the third board cut in the dining room. You can go to the dining room immediately and set the board alongside the third wall from where you started measuring in that room.

7. Nail 'em up—When all the pieces are lying in front of their respective locations, it is time to crawl around the room and nail them up. You may occasionally find that a board or two will need to be recut or trimmed because of walls being out of square or because you measured wrong. This would happen regardless of your approach to the job and will not affect your other cuts. You can either recut as you come to them or write the adjustment needed on the back of the board; e. g., "-⅛" indicates that you need to cut ⅛ in. off of the length.

This system also works well if you like working with a partner. One person measures and installs, the other does all of the cutting. Work this way with a partner only if you like and respect each other and if each of you has a sense of humor. Inevitably there will be discussions about who can't measure right and who can't cut right when boards occasionally come up short or long.

Nothing fancy, but it works—It's a simple system, really, but it does work. You don't have to use a preprinted form, of course, though it does eliminate the need to write out the sequential numbers and draw lines. Besides, it's a heck of a lot easier to write on and read than some of the things that you see carpenters writing on at construction sites.

To save you a bit of setup time, you can photocopy the sample form at right for your own use. Photocopy enough to keep yourself supplied for a few months and keep a few extras in your truck. The columns will allow you to work through the "measure, cut, nail" process 60 times per page. You can also use the form to record measurements for other types of molding and for closet poles. Baseboarding may still be the lowliest job going, but with a bit of organization, you won't be down there quite as long. ☐

Greg Smith is a general contractor in West Los Angeles, Calif., who specializes in custom home building and remodeling. Photos by Marilyn Ray.

Cut angles diagram labels: 45° miters (two outside); 22½° miters (one inside, one outside); 45° miters (two outside); Walls; 45° miter (two inside); 22½° miters (one inside, one outside); 45° miters (two inside); Drywall; Square cut; Casing; Jamb

Cut angles
These cuts are commonly needed when running baseboard. The drawing shows mitered corners, but some carpenters prefer to cope the corners. The author's measuring system will work in either case.

Drawings: Bob Goodfellow

Molding Character

Using a molder/planer on site to create a formal 18th-century interior

by Douglas Honychurch

As a Connecticut native and real-estate appraiser, I've gradually come to appreciate the examples of early Colonial architecture that remain in our area, and I particularly value the craftsmanship that made them worth preserving. My interest began with old houses that I noticed while driving through the New England countryside, and eventually led me to look for historic houses that were open to the public. I began buying books about Colonial houses and enjoyed studying their wonderful architectural details. Some of the books had cross-sectional drawings of walls, showing wainscoting and crown molds. These especially interested me.

The desire to do some of this work myself led me to purchase a Williams & Hussey molder/planer, and use it to run moldings for a remodeling project in my own house. But my real chance to design and mill Colonial trim came when my brother asked me to help with a new house that he was building. I didn't need much coaxing. The house is a two-story Dutch Colonial with flared eaves, cedar clapboards and a corbeled end chimney (top right photo, next page). My brother gave me free rein in the design and construction of the architectural details.

Design—With a slate-faced fireplace on one wall and a large, multi-pane window on another, the 420-sq ft. living room was a designer's paradise. We decided to build twin bookcases on each side of the large window. A raised-panel wainscot would enclose the room, with a two-piece chair rail above and a two-piece baseboard below. The door and window casing would have a beaded edge toward the opening and a small band mold around the outside. On the doors, this casing would die into a con-

toured plinth block at the floor. The formal mantel surrounding the fireplace would be flanked by fluted pilasters (photo below). There would also be an elaborate crown mold all the way around the room. For all this work, we wanted to forego stock moldings and make our own trim wherever possible.

The designs came primarily from three books: *Southern Interiors of Charleston, South Carolina* by Samuel and Narcissa Chamberlain (Hastings House Publishers, 1956), *Architectural Treasures of Early America,* vols. 3 and 7 (reprinted in 1977 by Arno Press Inc.) and *Early Domestic Architecture of Connecticut* by J. Frederick Kelly (Dover Publications, 1963). The first two books are out of print (I found them in used-book stores), but the third is still available.

My brother and I wanted to create a room that felt like the late 18th century. Decisions

about what moldings to use where were based on my taste. I liked the wainscoting and fireplace designs in *Architectural Treasures of Early America*. But they were too elaborate. We simplified the moldings and eliminated the carvings. These decisions were also influenced by the limitations of my molder/planer, which can cut only ¾ in. deep. And of course, trial and error played a part in the design—tack on a molding, stand back and look—a process that led to many changes.

Construction—Unfortunately, the carpenter who had done such a fine job framing the house was behind schedule for his next job and couldn't be persuaded to do the complex trim work on the first floor. Our search for an equally skilled carpenter took some time, but on the recommendation of our painter, we met Ed Rockwell and looked at some of his recent work. We soon learned how fortunate we were to find him. His experience with this type of finish carpentry was extensive, since his family had been in the business for several generations. Unlike some tradesmen who are steeped in traditional lore, Rockwell was willing to try new designs and methods.

With our carpenter on the job, we were ready to make the moldings. The molder/planer we used was a Williams & Hussey model W7S (photo right), with power infeed and outfeed rollers (Williams & Hussey Machine Co., Elm St., Dept. 1361P, Milford, N. H. 03055). When I got the machine in 1984, I also bought a 2-hp motor to go with it. The whole outfit cost about $800.

We set up the molder/planer in the living room alongside Ed Rockwell's table saw, which was mounted on a rolling stand with a plywood extension added to the back of the table, both of which I found quite helpful. We used kiln-dried, select eastern white pine from New Hampshire. Ultimately we made seven different moldings (drawing, bottom right) and used three of them in the living room. The first step was to rip the boards to width on the table saw. We used a high-quality carbide-tipped blade, which made cleaning up the edges unnecessary.

Williams & Hussey offers a range of molding knives in standard patterns, and the company will also custom-grind special patterns. All you have to do is send them a drawing or a sample of the molding you want to cut. The cost of the knives is based on the overall width of the high-speed steel blanks (from 1 in. to 7 in.), not on the profile of the molding. I had custom sets of knives made, and they ranged in price from $72 to $120 per set.

To install these knives (two per set) on the molder/planer required only tightening some bolts. The alignment and registration of the knives were perfect without any adjustment. We used two plywood fences on the molder/planer, and secured them with small C-clamps that come with the machine. The knives had to be lowered into position and the fences adjusted and positioned so that the dimensioned boards would run through without any slop. We put the board between the two fences and lowered the knives to see where they would hit. If the alignment was right, we next adjusted the tension on the

Surrounded by fluted pilasters and a pair of raised panels, the fireplace mantel (photos above left and previous page) commands attention in a room full of architectural details. Both site-made and stock moldings were used for the trim work. To cut costs and avoid problems with wood movement due to moisture, the raised panels were made in one piece, using medium-density particleboard. The interior detailing, inspired by that of the 18th century, was a natural choice to complement the Dutch-Colonial styling of the house (above right).

Using a Williams & Hussey model W7S molder/planer on the site, Honychurch made many of the moldings used to trim the first-floor rooms. Molding profiles are shown in the drawing below.

Moldings made on site

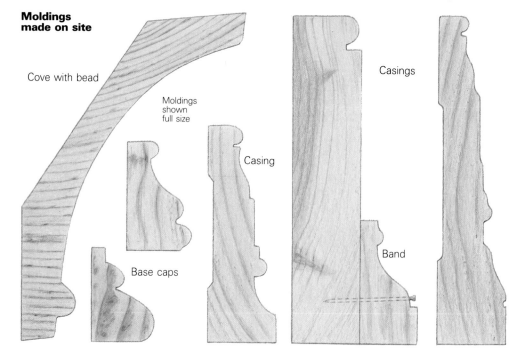

Cove with bead

Moldings shown full size

Casings

Casing

Band

Base caps

The mockup of the Greek-key fretwork on the mantel (above) was rejected because it would have been too complicated to make and apply. Saw kerfs were made in the back of the pine board to relieve cupping and were later covered by a thin band.

The crown mold in the living room (above) is just one 5-in. wide piece, though it looks built up from three pieces. It's a discontinued stock design called Curtis mold, and was ordered from a local mill. Around the fireplace (below), beaded cove mold was added under the crown.

feed rollers, which pressed the wood firmly onto the table while cutting, in addition to forcing it through the cutterhead. We learned how important feed-roller tension was when we broke a blade because the rollers weren't tensioned enough, and the stock began to chatter. Adjusting them was a real challenge—too little tension and the stock would flap violently and get chewed up; too much and it wouldn't feed.

Overall, the machine performed well for our needs and allowed us to make custom molding at less than custom prices. The blades stayed sharp; we used one set to make door and window casing for the entire house. These did need sharpening at one point, for which Williams & Hussey charged $15. To keep the knives sharp, it was important to prevent resin, pitch and sap from building up on them. We did this by carefully scraping them with an old chisel.

The fireplace—In the center of the 30-ft. long exterior wall is the fireplace. We wanted it to be the focal point of the living room. So one of our first decisions was to extend the 8-ft. section of wall that encompassed the fireplace into the room about 4 in., thereby breaking up the long expanse. This 8-ft. section is flanked by fluted pilasters, which we made on the job using a router and a fluting bit. The area between the pilasters is finished with an elaborate mantel and two large raised panels above it.

Our mantel design was suggested by one that we found in *Architectural Treasures of Early America*, vol. 7, and was a popular style in the eastern United States during the late 1700s. In deference to cost, we simplified the design by omitting the carved elements and using stock moldings to supplement the ones we made ourselves. We built the mantel directly onto the wall, furring out the main section with 2x4s nailed flat and covered with ¾-in. plywood to provide plenty of nailing surface.

At one point in the course of my trial-and-error design for the mantel, I considered using a Greek-key fretwork under the mantel shelf. I experimented on my table saw, cutting alternating grooves in a 1-in. piece of pine (photo top left), but realized that running it around the various offsets on the mantel would be tough. At this point, a trim carpenter who was working upstairs walked by and saw what I was trying to do. He didn't like the molding I'd made or how difficult it would be to apply, even though it wouldn't even be his job. He told me "You really know how to try a man's patience." He then built a jig for his router and made an alternative dentil molding that was good-looking and simpler to apply. It's the one we used.

Rockwell suggested that we embellish the crown mold in this 8-ft. section, and so we added a beaded cove at the bottom (photo bottom left). Also, at the extreme outside corners of this section, rather than a butt or mitered joint, we used a nearly circular beaded joint. Altogether, it took Rockwell and me a full week to complete the mantel.

Wainscoting—We decided to run a raised-panel wainscot around the entire living room, dining room and hall. The 34-in. chair-rail height

After some debate over whether the chair rail should protrude beyond the casing, the detail above was chosen as the appropriate way to end the molding at doors and windows.

was worked out to accommodate panels of pleasing proportion, separated by 3½-in. wide rails (horizontal border) and stiles (vertical border). In laying out each wall and determining panel sizes, we tried to keep the room as symmetrical as possible.

We chose not to use the traditional method of constructing raised panels—with stiles and rails mortised together, the raised panel let into a groove within them and the finished frames and panels applied to the wall. We wanted a method that would be less costly. Before sheetrocking, we had nailed plenty of blocking between all the 2x4 studs at the anticipated heights.

We started with the rails, nailing 1x6 pine around the room at chair-rail height. Then we ran the bottom rail so that once the 1x6 baseboard and base cap were installed, 3½ in. would be exposed above it. After the rails were up, we cut the stiles to fit between them. We used butt joints with a little glue and just nailed the pieces to the wall. No half-lapping was done. The pine wasn't a uniform thickness so we had to shim some of the joints to make them flush.

We then measured all the openings and calculated how much material we'd need for the raised panels. To save time and money, we decided to experiment with a medium-density particleboard called Medite (Medford Corp., Medite Division, P.O. Box 550, Medford, Ore. 97501) to make the raised panels. It is extremely dense and smooth and takes a beveled edge well, in addition to being very stable. Medite is available in 4x8 (actually 49-in. by 97-in.) and 5x8 (61-in. by 97-in.) sheets and a range of thicknesses from ¼ in to 1 in.

By using sheet goods for the panels, we were able to cut each one from a solid piece, as opposed to traditional panels, which are usually made from glued-up pine boards. Also, by using Medite, which is moisture resistant, we were able to reduce the chance that the panels would warp or check.

Using a table saw equipped with a carbide-tipped blade, we cut the Medite into panels to fit the 63 openings. We bought a carbide-tipped

Traditionally styled bookcases flank the large, multi-pane window. Before they were attached, all

raised-panel cutter, and arranged with a local woodworking shop to use their large shaper (1¼-in. spindle) to bevel the panels' edges.

We were extremely pleased with the way the raised panels turned out. We took the numbered panels back to the house and found that most of them fit without any need for adjustment because we had intentionally undersized them ¹⁄₁₆ in. all around. Then, using brads inserted with a brad driver, we ran ⅜-in. quarter-round around the openings to hold the panels in place and cover the gaps. We later discovered another advantage of our paneling method. When the electrician needed access at a particular spot behind

the paneling, we were able to remove only that one panel with ease.

The chair rail, instead of being flat on top, has a slight wave in it, which then steps down before protruding out in a rounded edge. Its downscaled profile resembles a traditional 18th-century chair rail. This was one of the few moldings we ordered from a local mill. We ripped the chair rail to a width of about 3 in., then scribed it to the wall with a pencil compass. We sat it directly on top of the 1x6 and nailed it into the studs. Underneath it, we ran the same band molding as on the door and window casings.

Rockwell and I disagreed about the treatment

the facings were beaded along their inside edges—a simple detail that adds depth and richness.

The same wainscoting design used in the living room also covers the walls of the foyer and runs up the stairs. But the crown molding in this area is a simple beaded design and was made on the site.

of the chair rail where it met the door and window casing. As I had designed it, the chair rail protruded about 1 in. beyond the casing. Rockwell strongly objected to this, believing that the profile of the chair rail should be shallow enough to die into the casing without sticking out. I insisted on my design and he relented, notching the chair rail so that it lapped onto the casing as much as it protruded beyond it. When he applied the band mold to the door casing, he decided to cope it around the chair rail. He did this 26 times throughout the house, and even he seemed pleased with the result (photo facing page, left).

Crown molding—The crown molding we used in the living room is actually one piece but looks like it was built up with at least three pieces (top right photo, p. 39). It is a discontinued stock design called Curtis molding and is made from a single piece of 5/4 by 5-in. pine. This molding has been made on the Williams & Hussey molder/planer, but we felt that it would be pushing the limits of the machine, and we therefore had it made for us at a mill.

For the crown molding in the center hall (photo above right) and dining room, we used a 4-in. beaded cove that I designed. This we did make on our machine. The profile required a

¾-in. deep cut, which the manufacturer says is the maximum. We made the molding in one pass, and it did test the machine's limits. We could hear the motor strain; in fact, the power demanded to cut this molding was so great that the lights dimmed when we ran it. A much simpler molding than the Curtis mold, it was easier to install, especially when coping the inside corners.

Bookcases—The bookshelves and cabinets that flank the multi-pane window (photo left) were designed by Rockwell, based on ones he'd done in the past. He built them in place, with 10-in. deep adjustable shelves above and 16-in. deep cabinets below. The upper portion of each bookcase has two bays made from 1x10 pine. Rockwell beaded the inside edge of the facings with his router. The exposed parts of the base were made with birch plywood, including the counter, which was then nosed with window stool ripped to 1¼ in. Under the nosing we ran the same band mold as below the chair rail. Rockwell made the raised-panel cabinet doors on site, using his router and table saw. The edges of the door openings are also beaded, which adds a satisfying depth to the design. □

Douglas Honychurch is a real-estate appraiser in Trumbull, Conn.

Making Plaster Molding

Sometimes complex shapes and profiles are easier to make in plaster than in wood

by Frank C. Freyvogel

When my wife and I added on to our modest-size Cape, we installed three exterior doors, each of which had an elliptical transom. We both love the look of these transoms, but trimming out these exotic shapes can be difficult, even to an experienced contractor such as myself.

We needed more than 50 ft. of door trim, including the straight sections and ellipses. My first choice was to have the trim custom-fabricated, but the prices I was quoted were astronomical. Next, I considered making the trim myself. I've had some experience laminating wood for odd-shaped applications. But between the cost, the set up and the lengthy time involved, I decided against this alternative.

Looking for another solution, I realized that much of the interior-trim work I've seen over the years was done in plaster. But most of my plastering experience, beyond asking questions and

reading, has been limited to patching. Nevertheless, armed with reference books, limited experience and lots of motivation from wife and wallet, I decided to trim my transoms in plaster (photo above).

Making the molding template—The basic process of making plaster molding involves building up layers of plaster where trim is going and dragging a molding template across each layer until the shape of the trim is created. My first step was to fashion a molding template out of steel stiff enough not to need the wood backing that normally would give the metal the necessary rigidity (top photo, facing page). I used 14-ga. steel I had on hand, but I recommend thinner steel, perhaps an old taping knife, because heavier-gauge steel is more likely to transmit file marks to the plaster.

The key to using the molding template successfully is the leg that rides on the jamb (bottom photo, p. 45). This leg ensures proper positioning of the plaster molding and determines the size of the reveal. I started with a piece of steel a couple of inches larger in each direction than the molding I was matching. I positioned a scrap of this molding on the steel and scribed the shape so that a 1-in. by ⅝-in. leg was left with a ³⁄₁₆-in. reveal. I cut the rough shape in the steel with a jigsaw, staying back a little from my scribed line. I used fine files and the grinding stone on a Dremel tool to shape the mold. When I was satisfied, I added a wood handle to make the tool easier to use and to protect the jamb from damage from the metal leg.

Prepping the walls—I cut back the drywall at the edge of the door jamb to provide a keyway

Photo this page: Roe A. Osborn

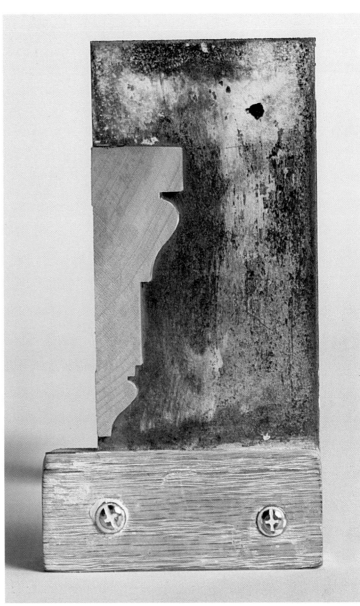

The molding template. The template was shaped to match the molding in the rest of the house. The author added a wooden handle covering the leg to make the tool easier to use and to protect the jamb from being scratched. The trim (left photo) is fashioned of plaster using the template (above).

for the plaster and to give the molding more body (bottom photo, right). I then primed the drywall and the jamb with two coats of primer to protect the paper face of the drywall from tearout due to the plaster's wetness. Next, I ran my molding template along the jamb and penciled in the outer edge of the new trim. The area inside my line was covered with a bonding agent that allows plaster to adhere chemically to a subsurface. The bonding agent also kept the plaster from drying too rapidly.

In retrospect I should have applied the bonding agent 3 in. or 4 in. beyond the outside boundary of the trim because there was a tendency for a small amount of excess plaster to build up beyond the trim and collect in irregularities in the wall. Unless this excess plaster is cleaned as you work, which you will have little time for, it eventually flakes off, creating future headaches.

A keyway locks the molding in place. The wallboard has been cut back from the door jamb, creating a void that fills with plaster and keeps the molding securely anchored.

The first layer of plaster is put on. After the primer and bonding agent have been applied to the molding area, the author uses a taping knife to apply a stiff mixture of plaster. The molding template then is dragged over this first layer of plaster, and the basic shape of the trim is created.

Although I skipped this next step, I strongly recommend the addition of expanded galvanized metal lath mechanically fastened to the jamb and to the wall to strengthen the trim.

The final step in preparation was protecting my work area with plastic, 4 mil or better (especially if you are working over hardwood floors—lime is caustic and will discolor hardwood).

Mix the lime putty ahead of time—In addition to my molding template, I needed other equipment, including a hawk (a small metal board on a handle used to hold the plaster while working it); a trowel; and a 6-in. taping knife, all available from the local mason-supply yard. I also picked up a bag of hydrated finish lime, a bag of molding plaster (or gauging plaster) and a small quantity of commercial retarder.

I made the lime putty the day before I needed it. This allowed the lime to soak up the water

evenly and helped to eliminate lumps in the mix. I filled a 5-gal. pail about three-fourths full of clean water (14 qt. per 25 lb. of lime) and added the lime a little at a time. I stirred the mix occasionally as I added the lime, and I let it sit overnight. The next day, I stirred it again with a stick (a heavy-duty drill and paddle also would work well). The consistency of this lime putty should be almost the same as it is for drywall joint compound. A mix that is too wet can cause small holes to develop in the finished molding, while a mixture that contains too little water can produce small, hard lumps that shrink and leave little voids in the finished molding.

Mixing the plaster—When I was ready to begin my trim work, I mixed the retarder in a bucket of clean, potable water. The recommended ratio is 24 oz. of retarder to 4 gal. of water. The exact amount of retarder varies based on your ability,

the size of the job and the ambient temperature. Clean water is a must. Water that has been contaminated with lime or plaster scrapings from tools accelerates a batch of plaster beyond a usable setting time.

Next, I made a ring of lime putty on a plywood scrap about 30-in. square and filled the ring with my water/retarder mix (the ring of lime putty creates a kind of mixing bowl). Then I added the molding plaster, sifting in a little at a time. I was aiming for a 1:1 proportion of plaster to lime. I let the plaster sit until all the water was absorbed and until there were no dry spots. Next, I mixed the lime putty and plaster together until they were blended thoroughly. This initial mix of putty and plaster should be somewhat stiff so that it stays put on the wall.

Applying the plaster—Using a taping knife, I applied the plaster to the area being trimmed (photo facing page). This step is called blocking out. Next, with the handle of the molding template riding on the inside of the jamb, I ran the template over the area I just blocked, working up from the top of the plinth block. It is important to keep the molding template perpendicular to the wall at all times to make sure that the shape of the molding is consistent.

The blocking operation created the basic shape of the trim. Once the initial blocking was done, I immediately went back and filled in any voids with the same material and passed the template over the area again. I repeated this process until the rough shape of the mold had been transferred to the plaster or until the material had become too stiff to use. *Never add water to soften your mixture!* Always discard any used material, and take the time to clean the excess plaster from your tools.

The next step required a looser mix (more water/retarder) and a smaller batch of plaster. I smeared a handful of the mix over the trim (top photo, right). Then I ran the molding template over that area, forcing the plaster into the voids that were left by the earlier blocking-out process. I repeated this procedure until all of the imperfections had disappeared.

Instead of smearing this layer of plaster, another option would have been to hold a handful of the loose plaster mix against the molding template as it passed over the trim. This operation is known as stuffing the mold. An extra person would be helpful with this activity because it normally takes two hands to hold the molding template in the correct position. In any case, you shouldn't continue the shaping process if the initial layer of plaster has swollen, causing the molding template to bind and chatter. Remember that plaster expands as it sets.

When I reached the top center of the transom, I repeated the entire process beginning at the bottom of the other side and blending the two sides at the top (bottom photo, right). I always worked against gravity. Working down would have left the trim full of cracks and voids.

The final step was splashing water on the trim with a mason's brush and running my mold template over the trim one last time to remove any excess material or drips. Some plasterers use a handful of lime putty at this point and stuff the mold to fill in any minor blemishes and to give the trim a shine. If it looks good after the initial stuffing, however, I suggest leaving it alone. After the plaster dried, I did final refinements with drywall joint compound and sandpaper.

The entire process took me around four hours per door. This was probably twice the time it would have taken a professional, but it was certainly quicker than the time to fabricate and install wooden moldings, and at a fraction of the cost. In addition to the time advantage, this process gave me moldings that exactly match the curve of the door because the molding template rode on the actual door jamb. □

Frank C. Freyvogel is a contractor who lives in North Bellmore, New York, and works on Long Island and in New York City. Photos by Terri Freyvogel except where noted.

A second layer is smeared on top. A loose mix is spread over the first layer of wet plaster. As the template is drawn over the molding, the loose plaster is forced into the voids left from the initial shaping to complete the profile of the trim.

Fine-tuning the trim. With the template leg riding on the jamb, the author makes the final pass, filling in the last of the voids as he blends the two sides together in the middle.

Cutting Crown Molding

Calculating miter and bevel angles so you can cut crown on compound-miter saws

by Stephen Nuding

A few years ago, I purchased an 8½-in. compound-miter saw. It was light and compact, but had the same capacity for cutting large crown moldings as a regular 10-in. miter saw. Remodeling Victorian homes, I install a lot of crown so this seemed to be the perfect power tool for me.

I eagerly brought the saw to the job and set the miter and bevel angles for 90° corners, as indicated by the instruction manual. When I cut my first lengths of crown, the joints weren't perfect, but I figured that the walls and ceiling weren't perfect either, so with a little shaving here and there I was in business.

The next crown molding I had to install, however, was a larger one, and when I cut it and held it up to the ceiling, I was looking at a pie-shaped gap ⅜ in. wide. What's more, this room had two corners that were 135°, not 90°, and the saw's instruction manual gave no miter or bevel angles for this situation. I soon discovered that throwing miscut pieces around the room in rage and frustration is a very slow and expensive way to complete a job.

By now I was ready to return the saw to the dealer and demand a refund. But in desperation I grabbed the instruction manual one last time. According to the manual, the miter and bevel angle settings were correct for 90° corners when using a standard crown, which

makes a 38° angle to the wall. Wait a minute… what if my crown doesn't make a 38° angle to the wall?

Fortunately, my daughter's protractor was in the car, so I was able to measure the angle the crown made to the wall by holding it against the inside of a framing square. The angle was more like 43° or 44°. I checked all the crowns I was installing only to find that none were the same, varying from 35° to 45°.

I finished the day's work as best I could and went home determined to calculate the angle settings for each of the crowns. Using my wife's high school math text to brush up on some trigonometry, I wrote down equations and measurements. I worked late into the night, but couldn't come up with a formula.

I finished the crown job eventually by trial and error, playing with the angles on the saw until they were right. Still, the problem gnawed at me. I spent a lot of late nights scribbling and thinking, but I just couldn't get it.

Fortunately, I had hung some French doors in the home of Roger Pinkham, professor of mathematics at the Stevens Institute of Technology. So one Saturday morning, at my request, he graciously came to the house and we pored over my notes. Several hours later, we had it. We could calculate the miter and bevel angles for any crown and for any angle.

So, why use a compound-miter saw?—You are probably wondering why anyone would want to calculate angle settings for a compound-miter cut when crown molding can easily be cut on a regular miter saw with no math at all. With a regular miter saw, the crown is positioned at an angle between the fence and the table (photo below left), but is turned upside down so that the wall face of the crown is against the saw's fence and the ceiling face of the crown lies on the saw's table. The crown is then cut at 45° to create a 90° corner, 22.5° for a 45° corner, and so on. Very simple. (For more on cutting crown molding, see the article on pp. 50-53.)

Most 10-in. miter saws, however, can only cut crown molding up to about 4½ in. wide. Five and one half-in. crowns are readily available, though, and cutting these requires a 14-in. or 15-in. miter saw—a large, heavy tool. Cutting large crowns on any of these saws also requires the extra step of constructing a jig or fence extension, preferably both.

Even a 15-in. miter saw is not big enough to cut crown molding more than 6½-in. wide, and larger crowns are also available. For instance, the Empire Molding Co., Inc. (721-733 Monroe St., Hoboken, N. J. 07030; 201-659-3222) makes an 8¾-in. crown that I often use.

So unless you want to make a king-size

To miter crown with a standard miter saw, turn the molding upside down and set it at an angle between the fence and table.

To miter crown molding with a compound-miter saw, lay the molding flat on the saw's table.

miter box and cut the molding with a handsaw, you'll have to use one of the new slide compound-miter saws or a radial-arm saw to cut these wide crown moldings (for more on these saws see *FHB #57*, pp. 58-62). With a compound-miter saw, crown molding is laid flat on the saw table (photo right, previous page). No jig or fence extension is necessary. The saws can be smaller for cutting the same size crown, resulting in a lighter tool with a smaller blade, which is therefore cheaper to buy and costs less to sharpen.

Figuring the angles—To calculate the miter and bevel angles for any crown molding, you'll need a framing square and a calculator that's capable of doing trigonometric calculations. These calculators are usually called "scientific calculators." No cause for alarm, though, just think of yourself as a carpentry scientist. I use a Radio Shack model that is out of production now, but a Radio Shack EC 4008 will do nicely and retails for only $13.95.

So, here we go. First let's consider the most common case, the 90° corner. Hold whatever crown molding you're using up to the inside of a framing square as in the drawing right. Measure lines A, D and C to the nearest 16th of an inch. (To convert fractions of an inch to decimals, simply divide the denominator into the numerator. To convert 7/8, for instance, divide 8 into 7 and you get .875.) The miter-table setting (M in our equation), is the inverse tangent of (A divided by C).

$$M = \tan^{-1} (A \div C)$$

To calculate this, divide A by C, and then hit the inverse tangent button (tan⁻¹), or arc tangent button (same thing). In our example, 2.875 (A) divided by 4.8125 (C) = .5974. With .5794 still on the calculator screen, hit the inverse tangent button and you get 30.9° (rounding to the nearest tenth of a degree). This is the miter angle at which to set your saw.

The bevel angle (B in our equation) is the inverse sine of D divided by (the square root of 2) times C.

$$B = \sin^{-1} \left(\frac{D}{\sqrt{2} \times C} \right)$$

To calculate this, multiply the square root of two (done on the calculator) times C. Then divide that into D and hit the inverse sine button, or arc sine (same thing) button on the calculator. Using the values from drawing A, the calculations would go like this: the square root of 2 = 1.41, times 4.8125 (C) equals 6.8059, divided into 3.875 (D) equals .5694, the inverse sine of which is 34.7°. This is the bevel angle at which to set your saw for a 90° corner.

Once you have calculated the miter and bevel angles for a particular molding, you never have to calculate them again as long as you have 90° corners. Jot down the angles somewhere and save a couple of minutes the next time you run that crown.

What if you have a wall corner that is not 90°? To make this calculation you'll need a device for measuring the angle of the wall corner. I use the Angle Devisor (manufactured by Leichtung Workshops, 4944 Commerce Parkway, Cleveland, Ohio 44128; 800-321-6840). Whether you are installing inside corners or outside corners, be sure to use the angle of the inside corner (the angle less than 180°) for the equation.

Here's how the equation looks:

$$M = \tan^{-1} \left(\frac{A}{C \times \tan (F \div 2)} \right)$$

If we were to use our crown from drawing A, we would have 135 (F) divided by 2 = 67.5. Hit the tangent button and you get 2.4142. That times 4.8125 (C) = 11.6184. Divide 11.6184 into 2.875 (A), then hit the inverse-tangent button, and you get 13.9° (the miter angle).

For the bevel angle:

$$B = \sin^{-1} \left(\frac{D \times \cos (F \div 2)}{C} \right)$$

Plugging in some real numbers we get: 135 (F) divided by 2 = 67.5, the cosine of which is .3827. Multiply .3827 times 3.875 (D) and you get 1.4829. Divide that by 4.8125 (C), then hit the inverse sine button, and 17.9 appears. That's your bevel angle.

Finally, because the difference of one degree in the miter angle or bevel angle can be the difference between acceptable and unacceptable joints, you must set the angles on your compound-miter saw carefully. Math may be perfect, but measurements and the real world aren't, so slight adjustments may be needed to get an acceptable joint. But by using these equations you will avoid the fuss-and-fiddle approach I first used. □

Stephen Nuding is a carpenter in Hoboken, New Jersey. Photos by Susan Kahn.

Figuring the angles

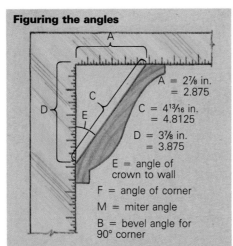

A = 2⅞ in.
= 2.875

C = 4¹³⁄₁₆ in.
= 4.8125

D = 3⅞ in.
= 3.875

E = angle of crown to wall

F = angle of corner

M = miter angle

B = bevel angle for 90° corner

Crown molding varies not only in size but also in the angle that it makes with the wall. So the first step in calculating miter and bevel angles is to measure the crown with a framing square and determine the measurements shown in the drawing above. Then plug those figures into the formulas shown below.

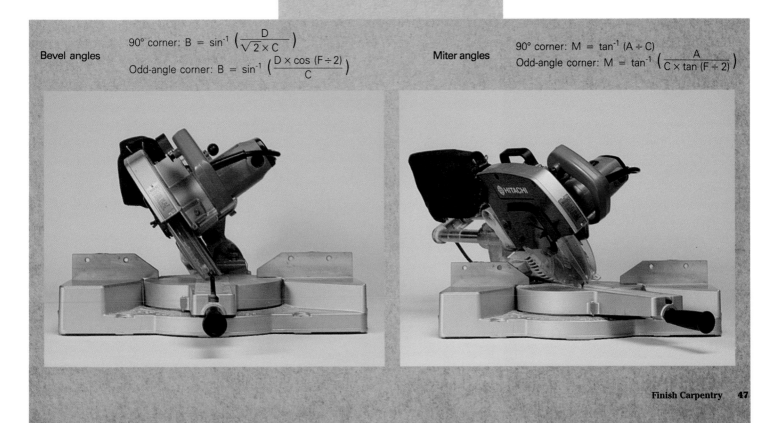

Bevel angles

90° corner: $B = \sin^{-1} \left(\frac{D}{\sqrt{2} \times C} \right)$

Odd-angle corner: $B = \sin^{-1} \left(\frac{D \times \cos (F \div 2)}{C} \right)$

Miter angles

90° corner: $M = \tan^{-1} (A \div C)$

Odd-angle corner: $M = \tan^{-1} \left(\frac{A}{C \times \tan (F \div 2)} \right)$

Table-Saw Molding

The secret is in the order of cuts

by Bruce Andrews

When the landmark Winooski Block was finished in 1862, the builders festooned it with all manner of ornamental moldings and wooden filigree. But by the time we (Moose Creek Restorations Ltd.) got the repair contract, 117 Vermont winters had weathered, cracked and split all of its remaining woodwork. Three-fourths of the building's cornice moldings were either rotten or missing. We were to replace 10,000 linear feet of various moldings, not one of them a type manufactured today, and we didn't even own the usual tool for milling moldings, the spindle shaper. We still were able to complete the job, relying on our table saw and a lot of careful planning. We found that the table saw could handle most any profile—it could even scoop out concave curves—but we also learned that every profile required its own sequence of cuts. Figuring out that sequence is the heart of our method.

The first thing we worried about was getting enough good stock. Molding stock must be the highest quality, close grained and knot free. We were still short of stock after several deals to obtain a couple thousand board feet of Vermont pine in varying widths, thicknesses and lengths—all rough cut and in need of finish planing, dimensioning, and in some cases, drying. We were bemoaning our plight when two young entrepreneurs wandered into our office. They asked if we knew anyone who could use several thousand board feet of redwood and cypress beer-vat staves from the old Rheingold Beer brewery that was being dismantled in Brooklyn, N.Y. Well, yes, we probably knew someone. The wood reeked of stale beer, but it was superb for our purposes. It was straight, close grained and of course, well seasoned.

Before any shaping could be done, we had to prepare our stock. We thought that the wood might have nails hidden in it, but we found none. We did find metal flecks where the vat bands had deteriorated, but with wire brushes and large paint scrapers we removed almost all the rust. On our 16-in. radial-arm saw, we ripped the lumber to the rough sizes we needed, about ¼ in. thicker and ½ in. wider than the dimensions of the finished moldings. Next we prepared the stock on a jointer and a thickness planer. Once we had dressed down the old surfaces ¼ in., the wood was perfect and unmarked. As we worked,

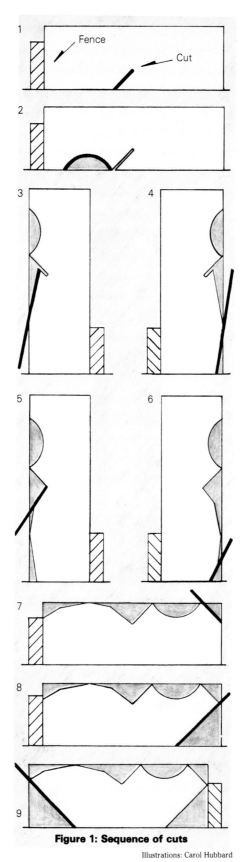

Figure 1: Sequence of cuts

Illustrations: Carol Hubbard

we checked our cutters for sharpness. Our stock was as straight and as square as we could make it; we were ready to begin shaping.

Setting up—Milling complex moldings on a table saw requires precision. Begin with an accurate template of the molding, to which you can adjust the sawblade's settings and against which you can compare results. The best template is a short piece of the molding you want to copy. If you must create a template from molding in place, you'll have to use a profile gauge. (See Figure 2 on the next page.) Many exterior moldings are too large to be handled with one application of the gauge. If this is the case with your trim, you'll have to take a series of readings, transfer them to paper and combine them for the complete profile. In fact, it's a good idea to sketch all molding profiles on site, for the gauges may get distorted before you return to the shop. Fashion your template out of a rigid material such as Masonite or plywood.

Before any cutting, even before setting the sawblade, scrutinize the template or molding cross section. The question is how to determine the order of the cuts. You don't want to take out a piece of stock you'll need later to run against the fence for making another cut. Think things through on a piece of paper. Certain cuts simply have to be made before others.

In Figure 1, for example, cut 1 is crucial because it is a dividing line between two curves: If its angle is incorrect or its cut misaligned, the proportions of both curves will suffer. If it is too deep, it undercuts the convex curve; if too shallow, material in the notch will have to be cleaned out later—a waste of time.

Cut 2, which creates the concave curve, must meet precisely the high point of cut 1. Because the stock is fed into the sawblade at an angle, this is a delicate cut.

Cuts 3 through 6, creating the convex curve, must be made after 2. If they had been made before 2, the convex curve would have made subsequent cuts a problem. (The stock could easily roll on that curve as it is fed into the sawblade.)

Cut 7 is delayed so that the point it creates with cut 2 won't be battered as the stock is maneuvered over the saw. Cuts 8 and 9 are made last, because leaving the corners of the stock square

The Winooski Block (left) is capped by a cornice assembly over 6 ft. wide; it consists of 14 elements, including 7 moldings that were reproduced on a table saw. The milling of the molding is described on the facing page.

You need a template to mill new moldings. Use a piece of the original, or transfer readings from profile gauge to paper on site and cut a template later in the shop.

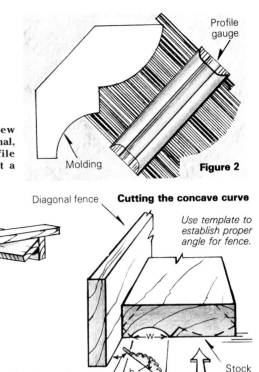

Profile gauge

Molding

Figure 2

Figure 3
Cutting setup

Straight 2x8 serving as diagonal fence

Fingerboards to hold stock down

10-in. C-clamps

Diagonal fence

Cutting the concave curve

Use template to establish proper angle for fence.

Stock

6 ft. to 7 ft.

Stock supported by board bolted to posts at exact height of saw table. Board braced at angle to direct stock.

Don't push stock across the blade at an angle greater than 60°.

Stock

Fingerboards to hold stock against fence

Concave curves may require several passes, starting with the blade set low. On the last cut, saw points should just touch curve outline.

ensures the stability and accuracy of preceding cuts. (Cuts 3 through 6 would have been almost impossible if cut 9 had preceded them.)

To save time, pass all molding stock through a given saw setting; be fastidious about such settings, making practice cuts on scrap work. Cut more molding stock than you'll need at each setting, so you'll always have waste stock with the necessary previous cuts. In other words, to get an accurate setting for cut 5, you'll need stock with cuts 1 through 4 already made.

Cutting—We used a 10-in. Rockwell Unisaw with a 48-point carbide-tipped blade for all molding cuts. For most cuts we used the rip fence provided by the manufacturer. For cut 2 however, we needed a diagonal fence, so we trued a 2x8, used a template to carefully set it at the proper angle for the desired cove, and clamped it to the table with 10-in. Jorgensen C-clamps (Figure 3). To reduce stock flutter we used fingerboards, pieces of wood with a series of parallel kerfs cut in one end. Two fingerboards clamped to the table held the stock against the fence, while one fingerboard clamped to the fence held the stock down. The kerfs allowed enough play to let the wood slide through, but maintained enough pressure to ensure a straight cut. Using fingerboards and extension tables, you could cut all the molding unassisted, but you may prefer to have a helper to pull the stock gently through the last few inches of a cut. Several times a day, wipe the tabletop and sawblade clean with turpentine, to minimize binding.

Except for cut 2, all cuts were made with the rip fence running parallel to the blade on one side or the other. As shown in Figure 1, cuts 1 and 2 were made with the stock face down on the table, while cuts 7, 8 and 9 were made with it face up. The stock stood on edge for cuts 3 through 5. (When cutting some symmetrical convex shapes, you can leave the sawblade at the same angle, and after one pass, turn the stock 180° to get the cut whose angle mirrors the first.) Each cut was preceded by carefully adjusting blade height and angle against the template. The last cuts on a molding (cuts 8 and 9) should be slightly larger than 45°—if your sawblade will tilt just a little more—to avoid gaps where the building surfaces are not quite perpendicular.

We cut the concave shape (cut 2) into the molding by passing the stock diagonally across the table-saw blade. (See Figure 3, at right.)To set the blade and fence correctly, you'll need a piece of the old molding. (A template is less effective.) Holding the high point of the curve over the blade, slowly crank up the blade so that the tip of the highest tooth just grazes that curve's apex; lock the setting and try a few cuts. To create the width of the curve, angle the piece of molding until all the teeth of the exposed blade lightly touch the arc of the molding. Another person should snug the fence against the angled molding and then clamp the fence to the table while you hold the molding in place. You'll have to tinker a bit to get the exact angle you need.

You can create almost any symmetrical curve with this method. Pushing the stock across the blade at a wider angle will result in a wider curve. However, the widest angle at which we

would push wood across the blade is 60°; with wider angles, not enough of the sawteeth are gripping and the blade will bind. (I'm not sure why, but a 48-tooth carbide blade binds up less than an 82-tooth one. It may be that the chips clear more easily.) If the blade is binding, make several passes to get the curve, starting the blade low and cranking it up ¼ in. for each pass. Don't get so wrapped up in your calculations that you become careless. Keep fingers clear of the blade. The speed at which you feed the stock must be determined on the job: Too fast and the blade will bind, too slowly and the wood will burn. The greater the angle of feed, the more often you should clean the blade.

The quality of the wood greatly affects the complexity of cuts you can make. Hardwoods are more difficult to mill without proper equipment. If concave curves are possible at all on hardwood, you'll have to make many gradually increasing cuts; the angle of the stock to the blade will be limited. Fortunately the grain in our cypress varied less than ¼ in. in 15-ft.

To refine the shape of our convex curves, we used many tools, including jack planes, curved shavehooks and spokeshaves. Among power sanders, Rockwell's Speed-block was the favorite; we clamped the finished molding to benches and sanded it using 50-grit pads.

Using these techniques we milled 1,000 linear feet for each of nine molding types, some more complex than the one described above.   □

Bruce Andrews is a partner of Moose Creek Restorations Ltd., in Burlington, Vt.

Installing Crown Molding

Upside down and backwards is the secret

by Tom Law

The first piece of molding (left) is cut square and run into the corner. The second piece (right) is cut to the shape of the molding's profile (coped) and will butt neatly into the face of the other piece. The paper-thin point on the bottom of the coped piece will make the finished joint look like a miter.

The old-time carpenters I learned from used to amuse themselves by quizzing young apprentices about the trade. If you could answer the easy questions, the last question would always be: "How do you cut crown molding?" And when you looked puzzled, they'd go off, chuckling to themselves something about "upside down and backwards." Of all the different moldings, crown molding is the most difficult to install, largely because of how confusing it can be to cope an inside-corner joint.

In classical architecture, crown molding (sometimes called cornice molding) is the uppermost element in the cornice, literally crowning the frieze and architrave. These moldings were functional parts of the building exterior when the ancient Greeks used them, but they have been used for centuries on interiors purely as decoration.

Crown molding is installed at the intersection of the wall and ceiling. Originally crown molding was triangular in cross section—the portions abutting the wall and ceiling formed two sides of a right triangle, and the molded face was the hypotenuse. But only the molded face is visible, so much of the solid back has been eliminated to save material. Also, by eliminating part of the back, only two small portions of the molding bear on the wall and ceiling surfaces, which makes crown easier to fit to walls and ceilings that aren't straight or that don't form perfect right angles.

The crown tools—When I cut crown, I like to work right in the room where the molding will go so I can orient myself to the wall I'm working on. If a room is finished, however, I may have to do the cutting somewhere else. Then I have to imagine the molding in place when I'm positioning it in the miter box (and believe me, this can get tricky with crown molding).

I cut and install crown molding with hand tools. I use a wood miter box (top left photo, p. 52) because it's the kind I learned on, but also because my view is not obstructed by the electric motor of a power miter box. Installing crown molding is slow and calls for careful work, so the production speed of an electric miter box is not required. I cut miters with a standard 26-in. handsaw (10 or 11 point). Miter cuts are made through the face of the molding, and a sharp handsaw will do a better job than a dull circular-saw blade will do.

For this kind of work, I prefer a workbench to a sawhorse. Mine is just a simple frame of 2x4s and 1x4s with a 2x12 top. It stands 34 in. high, which is a more convenient height to work on than a sawhorse provides. You don't need to deliver a lot of power to cut trim. A broad bench top is also convenient for holding tools.

Although you can still find deep-throated coping saws in the mail-order catalogs, most coping saws nowadays are 5 in. deep and have a 6-in. blade. The blades come with different numbers of teeth. I try each kind of blade to see which works best with the wood I'm cutting. Generally, finer teeth work best with hardwood and coarse teeth with soft wood, but not always. For the job shown here, I used a fine-tooth blade to cut soft wood. It works more slowly, but makes a smooth cut.

The blade of a coping saw can be inserted with the teeth directed toward you to cut on the pull stroke, or away from you to cut on the push stroke. Although it's strictly a matter of personal preference, I orient mine to cut on the push stroke because it acts in the same manner as a handsaw.

Measuring and marking—Crown molding can be used by itself or combined with other moldings, but it should always be in proportion to the size and height of the room in which it's installed. Too much molding at the ceiling line tends to lower the ceiling visually. Three or four inches of molding at the ceiling line is about right for an average-size room.

In the rooms shown in this article, I used 3⅝-in. crown molding, which is the most common size available. This dimension is the total width of the molding, but it's not the critical dimension that you use when installing crown. You need to know the distance from the intersection of the wall and ceiling to the front of the molding, measured along the ceiling. Because the back part of the molding has been eliminated, you can't measure this directly. Instead, I put the molding inside a framing square to form a triangle and read the distance (top drawing, right). The molding shown here measures 2⅟₁₆ in. I mark that distance on the ceiling at each corner of the room and in several places along the walls. These marks will serve as a guide when I install the molding.

It's frustrating to drive a nail through a piece of molding and not hit anything more solid than drywall, so I locate the framing members ahead of time—when I can still make probe holes in the wall that will be hidden by the molding. If the room hasn't been painted, you can spot the studs and ceiling joists from the lines of joint compound and make pencil marks on the wall to guide the nailing. If the room has been painted, you can find the studs and joists by tapping with a hammer and testing with a nail. Electrical outlets and switches are nailed into the sides of studs and offer a clue to stud locations.

Running crown molding should be one of the final jobs on a new house. Walls and molding should be primed and first-coated, the molding installed and then finish coats applied. If you're retrofitting crown molding, it should be prefinished entirely so that all you need to do is touch up the paint or stain after installation.

On this job the walls were painted and the molding was prestained, so I marked the stud locations right on the crown molding. Rather than use a pencil to mark the wood, I made a slight hole with the point of a nail, which was easier to find in the dark stain and which I later nailed through.

Getting started—When I run crown molding in a typical room—four walls, no outside corners—I usually start with the

wall opposite the door (bottom drawing). Unless it's perfect, a coped joint looks better from one side (looking toward the piece that was butted) than it does from the other (looking toward the piece that was coped). By first installing the crown molding on the wall opposite the door, and coping the molding into it on both ends, the two most visible joints show their best side to anyone entering the room.

I put the first piece up full length with square cuts on both ends. I cut it for a close fit, but if it's a little short I don't worry. Any small gap will be covered by the coped end of the intersecting piece. I hold the molding in place, lining up the top edge on the 2⅟₁₆-in. mark, and nail it. I use the shortest finish nails that will reach the framing, usually 6d or 8d. I nail into the wall studs through the flat section of the molding near the bottom, and I nail into the ceiling joists or blocking through the end of the curve near the top. I don't nail too close to the ends when I first put the piece up. I leave them loose to allow for a little alignment with the intersecting piece.

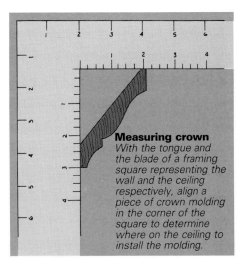

Measuring crown
With the tongue and the blade of a framing square representing the wall and the ceiling respectively, align a piece of crown molding in the corner of the square to determine where on the ceiling to install the molding.

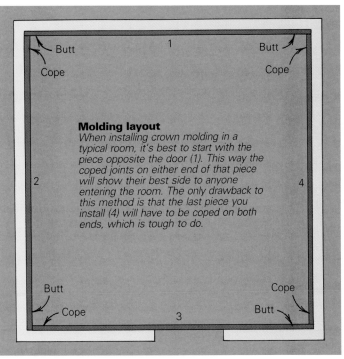

Molding layout
When installing crown molding in a typical room, it's best to start with the piece opposite the door (1). This way the coped joints on either end of that piece will show their best side to anyone entering the room. The only drawback to this method is that the last piece you install (4) will have to be coped on both ends, which is tough to do.

Occasionally I need to pull the top edge of the crown tight against the ceiling, but there's no ceiling joist or blocking to nail into. When that happens, I use 16d finish nails to reach all the way to the double top plate on the wall. Or sometimes I put a little glue on the molding and drive a pair of 6d finish nails at converging angles into the drywall about ½ in. apart (drawing next page). This pins the molding to the drywall while the glue dries. Another trick for nailing up crown when you don't have adequate framing is to nail up triangular blocks as shown in the drawing on p. 53.

I usually work around the room from right to left because I'm right-handed, and making coped joints on the right is a little easier than on the left. The second piece of crown to go up needs to be coped on the right and square cut on the left.

Coping with crown—Finish carpentry can be harder at times than cabinet work because you often have to make perfect joints against imperfect surfaces. Coping inside joints rather than mitering them is one way to deal with that problem. If you miter the inside corner with crown molding, the joint will often open up when you nail the pieces because the wall gives a little. A coped joint on crown molding won't open up and will be tight even when the walls are not exactly 90° to each other.

A coped joint is like a butt joint, with one piece cut to fit the profile of the other (photo facing page). The first piece of molding is cut square and run into the corner. The second, or coped piece is made by cutting a compound miter on the end to expose the profile of the molding, then sawing away the back part of the stock with a coping saw, leaving only the profile. The end of the coped piece will then butt neatly into the face of the first piece.

Because I don't always get the cope right the first time, I start with a piece of molding longer than I need and cope the end before cutting it to length. The phrase "upside down and backwards" refers to the position of the crown molding in the miter box when you're coping it (photo next page, top left). Crown molding isn't laid flat against the side or bottom of the miter box; it's propped at an angle between the two, just as it will be when installed. But the edge that will go against the ceiling is placed on the bottom of the miter box and is therefore "upside down." The right-hand side (if that's the coped end) is placed on the left and is "backwards."

The crown has to be positioned so that the narrow flat sections on the back of the molding, which will bear against the wall and ceiling, are square against the side and bottom of the miter box. When they are, the bottom should measure out the

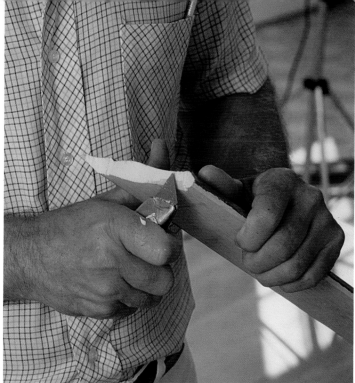

Before the crown molding can be coped, the end must be mitered to expose the profile. Positioned "upside down and backwards" in the miter box, the molding rests against small nails that hold it at the proper angle (top left). After exposing the profile of the crown molding, the back part of the stock is cut away with a coping saw, which must be held at a severe angle or the coped joint will not be tight (left). Even with the coping saw cutting at a severe angle, it's tough to remove enough wood through the S-curve in crown molding. Additional stock often has to be pared away with a utility knife (above).

required 2$\frac{1}{16}$ in. Once I find this position, I usually draw a pencil line on the bottom of the miter box to help me position subsequent pieces. Sometimes I'll even put a few nails on the line or glue a strip of wood to it.

Even with the molding positioned correctly in the miter box, it's still easy to cut it wrong. When I make the 45° cut to expose the profile of the molding for the cope, I remind myself that I want to cut the piece so the end grain will be visible to me as I look at it in the miter box (photo top left).

Coped joints are always undercut slightly, but crown molding has to be heavily undercut through the S-curve portion of the crown (called the *cyma recta)* or it will not fit right. I start the cope at the top of the molding using light, controlled push strokes. If I'm having trouble going from the straight cut to the curve, I back the saw out and come in at a different angle to cut away the waste. I begin the curved line with a heavy undercut and hold this angle all way through. I cut as close to the profile line as I can (photo above left).

The bottom of the crown molding is made up of a horizontal flat section, a cove and a vertical flat section. I cut down to the upper flat and then take the saw out and start cutting from the bottom. Some carpenters simply square off the bottom, but I try to leave the little triangular piece intact (photo, p. 50). I support it with my thumb as I'm coping and slice it paper thin. This little piece makes the coped joint look like a miter and helps close any small gap if the first piece didn't fit tightly to the wall.

I always test the cope against a scrap piece of molding to make sure I'm in the ballpark before actually trying it in place. Despite my best efforts to undercut the curved section, I usually have to pare away some more wood with my utility knife (photo top right).

I cut the piece just a little long and test it in place before cutting it to final length. If the fit of the coped joint is close, but still a little off, I can sometimes improve the fit by twisting both pieces either up or down the wall at this point—the 2$\frac{1}{16}$-in. mark on the ceiling isn't sacred. The buildup of spackle or plaster in corners can distort the intersection of wall and ceiling. Some carpenters carry a small half-round file with them to fine tune the fit of the cope.

Around the room—Once the coped joint fits, it's time to cut the piece to length. You can measure the total distance from wall to

wall, but I find it easier to measure from either of the two vertical flat sections on the molding that the coped piece will butt into. If I'm working alone, I either step off the measurement with a measuring stick (a 12-ft. ripping, for instance), or I'll drive a nail into the wall (above the line of the crown molding) and hook the end of my tape measure over it. Wherever I measure from on the wall, I'm careful to measure to the same place on the piece I'm cutting.

When the coped piece is cut to length, I nail it up just like the first piece, leaving the square-cut end unnailed for the time being. If I need to draw the coped joint tighter, I nail through the coped piece into the piece it abuts.

The third piece of crown molding goes up just like the second, but the fourth one needs to be coped on both ends (bottom drawing previous page), assuming the wall is short enough to be covered with a single piece of molding. I cut this piece about $\frac{1}{16}$ in. longer than the actual measurement, bow out the middle, fit the ends and snap it into place. The extra length helps to close the joints.

Some carpenters don't like having to cope the last piece on both ends because there's very little margin for error. The way to avoid this goes all the way back to the first piece of crown molding that's installed. Rather than put up the first piece with square cuts on both ends, you can temporarily nail up a short piece of crown molding and cope the first piece into it (photo left, facing page). Then take down the short piece, work on around the room and slip the butt end of the

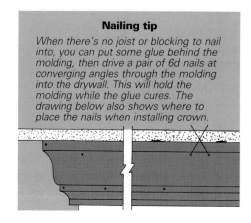

Nailing tip

When there's no joist or blocking to nail into, you can put some glue behind the molding, then drive a pair of 6d nails at converging angles through the molding into the drywall. This will hold the molding while the glue cures. The drawing below also shows where to place the nails when installing crown.

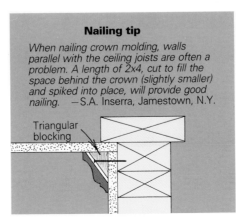

If you install the first piece of crown molding in a room by cutting both ends square, the last piece will have to be coped on both ends. To avoid this, you can put up a short piece temporarily and cope the first piece of crown molding into it (above). If an outside miter is open just slightly, sometimes you can close it by burnishing the corner with a nail set (top right). When a line of crown molding has to be neatly terminated on an open wall, the end should be mitered and "returned" into the wall with a small piece of molding. To avoid splitting such a delicate piece, it's best simply to glue it in place (right).

last piece behind the first cope that you made. This way all four pieces of crown molding in the room will have one square-cut end and one coped end.

When I go into a room that's not a simple rectangle, the decision about where to start is influenced by where I'll end. If there is an outside corner in the room, I like to end by installing the shortest piece that has an outside miter. That way, there's less wood wasted if I cut it too short. If there's not an outside corner, I like to work so that the last piece is installed on the longest wall that can still be done with a single length of molding.

When I need more than one piece to reach from corner to corner, I cut the moldings square and simply butt them together rather than use scarf joints or bevel joints. Butt joints are easier to make for one thing. And for another, although wood isn't supposed to shrink in length, the truth is it does. Over the years, I've seen a lot of joints that have opened up, and of those, the butt joints looked better than the others.

Outside corners—These are also mitered with the molding upside down and backwards in the miter box, but the saw is angled to bevel the piece in the opposite direction. When you miter for a cope, you expose the molding's end grain, but with a mitered outside corner, the end grain is *behind* the finished edge. Sometimes I cut them at an angle slightly greater than 45° to ensure that the outside edges mate perfectly. I usually add a little white glue, then nail through the miter, top and bottom, from both sides.

Sometimes outside corners will close tightly but the leading edge of one piece overhangs the other, perhaps because the corner is not exactly 90° or because one piece of molding is thicker than the other (more about that in a minute). If the molding hasn't been painted or stained, I'll trim the overhanging edge with a sharp chisel and sand it. This actually leaves a narrow line of end grain exposed at the outside corner, but once the molding is stained or painted, the end grain isn't very obtrusive. There are times when the molding has the finish coat already on it, and I can't do this because it would expose raw wood. In that case, I use my nail set to burnish the projection smooth (photo top right).

On this house, I ran crown molding in the foyer and had to terminate the molding at the stairwell opening. I ran the molding through the dining room, turned the corner at the

stair and ended the molding with a return—a mitered piece that caps the end of the molding. To make a return, I simply cut a miter for an outside corner on the end of a scrap of molding, then lay the piece face down on the bottom of the miter box and cut off the end. I glue this in place with white glue so as not to take a chance on splitting it by using a nail or brad (photo above).

What can go wrong?—Whether because the wood was wet when it was milled, or because the knives were dull, or because of internal stresses in the wood, the exact dimension and profile of the pieces in a given bundle of stock molding varies considerably. The differences aren't obvious until you try to fit an inside or an outside corner with two pieces that don't match. It's best to make joints from the same piece whenever possible.

There are times when the wall or ceiling is so crooked that gaps are left along the length of the crown. If there is a short hump that causes gaps on each side, I scribe the molding and plane it for a better fit. If the gaps aren't too bad, it may be best to fill them with caulk. Another trick I've used is that of leaving a small space (usually about ¼ in.) between the top of the molding and the ceiling, which makes it harder for the eye to pick up irregularities. If I'm doing this, I put up blocks to nail to, as shown in the drawing at left, and use a ¼-in. spacer block to ensure a uniform reveal. □

Consulting editor Tom Law is a carpenter and builder in Frizzellburg, Maryland.

Nailing tip

When nailing crown molding, walls parallel with the ceiling joists are often a problem. A length of 2x4, cut to fill the space behind the crown (slightly smaller) and spiked into place, will provide good nailing. —S.A. Inserra, Jamestown, N.Y.

Triangular blocking

Installing Two-Piece Crown

A method for running wide, paint-grade crown moldings

by Dale F. Mosher

I work as a finish carpenter on the San Francisco Peninsula, where there is a resurgent interest in formal houses that have a Renaissance European flavor. The houses often have a full complement of related molding profiles for base, casings and crown, and to be in scale with the rest of the building, these profiles can be quite wide. In the case of the crowns, I'm talking 10 in. to 12 in. wide. In fact, the crown moldings that I sometimes install are so wide they come in two pieces (photo at right).

There are several reasons for making crown in two sections. First, the machines that cut the moldings typically have an 8-in. maximum capacity. There is a lot of waste when wide moldings are carved out of a single piece of stock. For example, I'd need a 3x12 to mill a 10-in. wide piece of crown—an expensive, inefficient use of the resource. Two-piece crowns are also a little more forgiving during installation. The type I used on the job shown here can be overlapped in and out a bit, allowing the width of the crown to grow and shrink as needed to account for dips and wows in the walls and ceilings.

All the two-piece crown moldings I've encountered have been custom-made. The designer or architect comes up with section drawings; then the mill shop has the molding-cutter knives cut accordingly. Here, an average set of custom knives costs $35 per in., plus there is a $75 setup charge for each profile. So before the wood starts to pass over the cutters you've already spent a fair amount of money. But to create a certain look, it can be money well spent.

Stain-grade or paint-grade trim—If you've got one, your architect or designer will decide what grade the trim should be. If you don't, your checkbook will decide. At the mill, stain-grade means the stock is clear and virtually free of knots. On the wall, stain-grade means no opaque finishes will be applied. The moldings are individually scribe-fitted, and that

Joined at the shadowline. Wide crowns, which appear to be made of a single molding, can be made by running related profiles adjacent to one another. Any gaps between the two are caulked prior to painting.

Test fit. After affixing the top portion of the crown to its backing blocks, Mosher checks the fit of its corresponding bottom half.

means no caulks or putties to fill any gaps that may occur at miters and along uneven walls. Paint-grade material, on the other hand, will have some sapwood and uneven grain. On the wall, it can have caulkable gaps. Obviously, the stain-grade material will cost more—how much more depends on the species of wood. And it costs a *lot* more to install. Where I work, we figure four to five times more labor is needed to put up stain-grade crown as opposed to paint-grade.

The most commonly used paint-grade materials in these parts are alder, poplar and pine. I see more poplar than anything else because it's relatively inexpensive and easy to mill. Even paint-grade moldings, however, don't come cheap, and they should be handled with care. I have them primed on both sides as soon as they are delivered, and I store them on racks with supports no farther apart than 3 ft.

Miter-box station—I've used the new sliding compound-miter saws to cut crown, and I've decided to stick with my 15-in. Hitachi chopsaw. Here's why: When you're running crown, you've got to make both back cuts and bevel cuts. It takes time to adjust the saw back and forth, and the constant changes multiply the chances for error. Also, the crown has to lie flat on the table with a sliding saw, which makes it harder to see the path of the blade and the cut line. None of these is a problem when using a chopsaw.

The key to cutting crown accurately is having a good station for the miter box (top photo, next page). Mine has a pair of wing tables that flank the saw, connected by a ¾-in. MDF (medium density fiberboard) drop table that supports the saw. Each wing table is 6 ft. long and 2 ft. wide. They can easily be stood on end and carried through a standard doorway. I can also put a 2-ft. deep table against a wall and have enough clearance behind the saw to swing it through its settings.

The wing tables have fences that support backer plates for the crown during a cut. To install them, I begin by snapping a line on the floor where the assembled station will sit. Then I put the outside legs of the tables on the line, and anchor each leg to the subfloor. Next I put the miter box in the drop section of the table. I removed the stock fences from the miter box, allowing me to bring the arbor slightly forward of the original fence line, thereby increasing the width of the cut. The wider of these two crowns was 7¼ in.—just about the limit of what my saw can handle.

I make sure the saw's turntable can swing freely from side to side, and that its blade is square with the tables when set at 0°. I secure the saw to the drop table with four drywall screws run through holes that I've drilled in the saw's base. Each wing table has an 8-in. high fence that's square to the blade. I align the fences with a string to make sure they're straight.

During a cut, the crown bears against ¾-in. MDF backer plates that are screwed to the tables and the fences. The backer plates should be ½ in. wider than the widest crown section. The crown moldings for this job meet the wall and ceiling at 45°, so I ripped the backer-plate edges at 45°. Other crowns meet the wall and ceiling at different angles, and the backer-plate edges should be beveled accordingly.

It took me about a day to build this setup, and another half day to fine-tune everything. But the time it takes to build one will be returned tenfold in a single good-sized installation job.

Backing blocks—A two-piece crown needs a solid base for nailing and a flat surface to rest against to ensure correct alignment for the pieces. Backing blocks serve this purpose (top left photo, facing page). To find the width of the backing block, I assemble a couple of short sections of crown, as they would appear when installed, and measure the backside from the point the assembly hits the ceiling to the point it engages the wall. The backing blocks should be 1/16 in. less than this measurement to ensure that the crown will go together without leaving gaps between the pieces, at the ceiling, or at the wall. Backing blocks can be made of solid wood, but I prefer ¾-in. plywood because it's affordable, doesn't split and it holds nails well. After ripping a stack of backing-block stock, I cut the blanks into 6-in. to 8-in. lengths.

I prefer to place backing blocks on 16-in. centers, and no farther apart than 24 in. They should be affixed to the framing, so if the painting crew is about ready to prime the walls, I mark stud and joist locations with a

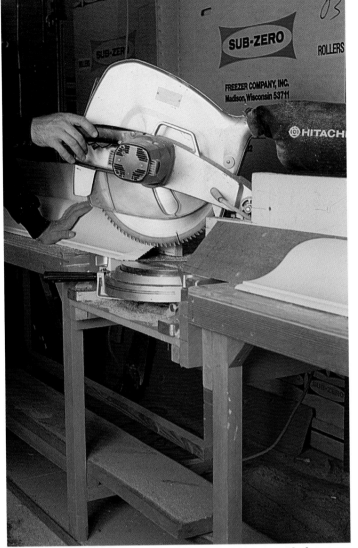

Miter-box station. **A pair of wing tables linked by a saw platform provides angled backer plates to support the crown moldings as they are cut. The plates are screwed to the table and fences.**

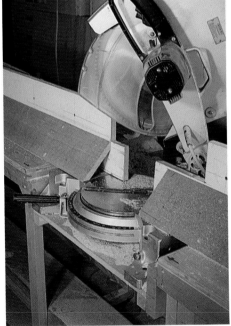

Dropped platform. **The wing tables are joined by a saw platform lowered far enough to bring the saw's table flush with the wings. The fences are braced from behind with triangular blocks on 1-ft. centers.**

keel (a carpenter's crayon) along the ceiling/wall intersection. A keel will bleed through most primers. Omission of this step gets you a one-way ticket to the planet of frustration, where you poke nails into the walls and ceilings, looking for the lumber.

Backing blocks are installed on layout lines snapped on the ceiling and wall. Using a torpedo level with a 45° bubble, I position a block at one end of the wall so the bubble reads level. The block should be about 2 ft. from the corner to avoid joint-compound buildup. I mark its edges on the ceiling and wall, and repeat the process at the opposite corner. These points are the registration marks for the chalkline. I use the raised chalkline as a straightedge to locate dips or bumps in the ceiling and walls. These problems are usually due to framing irregularities, and the backing blocks should be kept away from these places. I wish the framing crew could be around during this part of the job. If they only knew the trouble *we* go through to make *them* look good, they'd be taking trim carpenters to lunch a lot.

To attach the backing blocks I use a 2¼-in. finish nailer because finish-nail heads are small enough to be consistently set below the face of the block. Nail heads that stand proud interfere with the crown. When I've got framing on one end of the block for nailing, but none on the other, I put a bead of glue on the backing block to help anchor it to the wall.

At inside corners, I run one block into the corner, and then scribe the adjoining one to it. The backing blocks are typically a little too wide to fit between the lines in the corners because of joint compound on the wall, and they need to be trimmed a bit to fit. Outside corners are sometimes mitered as though crown molding, and secured to the wall, ceiling and to each other through their mitered edges.

When the backing blocks are up, I cut my "tester blocks." These are typically 16-in. to 24-in. long pieces of the crown molding. They need to be long enough to reach from an inside corner to the closest midspan backing block. I cut three pairs of tester blocks with inside miters at both ends, and three pairs with outside miters. One set has 44° miters at each end, one has 45° miters and the third has 46° miters. You may ask, "why not cope the inside corners?" For one, the curved profile of the widest molding in this job meant that a coped corner would be very fragile. I've found that a glued inside miter on a paint-grade job—if the pieces are carefully fitted—yields first-rate results.

Backing blocks. Crowns this wide need substantial backing to provide a consistent plane for aligning the two pieces and for adequate nailing. In the corner, one block extends to the wall while the other is scribe-fitted to it. Here, a test piece of crown is held in place. The pencil line along its point intersects the corner formed by the two blocks, marking the point from which the overall measurement for the crown will be taken.

Prybar tweaking. As the moldings are nailed home, a small prybar is useful for aligning the adjoining sections.

Unit assembly. Short sections of crown are best preassembled into a single piece. The pencil marks on the ceiling show the points from which the crown-length measurements were taken. On the right you can see a fully assembled run of crown.

Running crown—Installing crown is not a solo operation. The job will go a lot faster and with greater accuracy if you've got a good helper. Working on the theory that a piece of trim can always be made shorter, we begin with the longest run in the room by tucking the pair of 45° test blocks into one of the corners, just the way the finished crown will fit. If the fit isn't acceptable, we try a 44° and a 46° block until the right combination turns up. It might be a pair of 44s. It doesn't matter. It is very important, however, that the line of the miter line up with the corner, whether it's an inside or outside miter.

Once we find the best fit, we make a pencil mark along the bottom of the block into the corner (photos above). This marks the point from which the overall measurement is taken.

I don't bother to cut the piece a little long and then shorten it by degrees to ease into the fit. My helper and I can measure it accurately, so I cut it to that length. Period. This saves a lot of climbing up and down the A-frame scaf-

folds we typically erect as work platforms.

We also use tester blocks at outside miters to determine the best angles. To get our measuring points for an outside-to-outside miter, we make pencil marks on the ceiling to note the long points. For an outside-to-inside miter, we mark the long point of the outside miter, and the heel cut at the inside miter.

I'll typically put four 1½-in. finish nails into each backing block. The nails should be placed where the painter can easily putty the nail-heads. I don't put nails in a tight radius or too close to an inside corner. A small prybar can be useful for aligning the crowns during nailing (photo above left). I prefer the ones used by auto mechanics.

Back-beveling the miters can be useful on recalcitrant fits. A good tool for this is a 1⅛-in. belt sander. Its protruding belt makes it very maneuverable. If I need shims, I use pieces of manila folder. At each miter, I run a bead of yellow glue to ensure a sturdy joint.

Sometimes the crown has to work its way in

short sections around a wing wall. In this case, I usually preassemble the pieces if they're shorter than about 12 in. I put the parts together with glue and a pneumatic brad nailer, let the glue set for 20 minutes and then place it as a unit (photo above right).

As we run the upper crown, we make notes in the corners that describe any special angles or back-beveling that it took to get a good fit. Nine times out of ten, the same cuts will work on the lower section of crown.

After the crown is up, the drywallers can float on any necessary topping compound to hide the bumps and bows in the ceiling and wall. If the walls are to be textured, this should be done after the crown is installed. Our painters use oil-base putty to fill the nail holes, and latex-based paintable caulk to make the joint between the two pieces of crown disappear.□

Dale F. Mosher is a carpenter who specializes in finish work in Palo Alto, California. Photos by Charles Miller.

Making Curved Crown Molding

Glue-laminated trim can be shaped on a table saw

by John La Torre, Jr.

As a carpenter, I spend most of my time at work swinging a hammer or wielding a saw. But I'm always looking for a chance to try a new technique. Recently, while I was touring a house under construction, owner Paul Kreutzfeldt showed me a straight piece of stock crown molding he planned to use for the kitchen ceiling, then looked up at a curved wall in the room and said, "That piece is going to be a bear."

"Yup," I answered, excited. "Mind if I give it a try?"

"Go right ahead," he said with a smile.

There are two basic approaches to making curved trim. You can glue several pieces of wood end-to-end and then cut the curve on a bandsaw, or you can laminate thin strips of wood around a curved form. If you use the first method and decide to stain the trim, the separate pieces may accept stain differently, and the joints usually show through. Laminated trim, on the other hand, is stronger than butt-joined trim and usually looks better when it's stained. Because Kreutzfeldt had yet to choose between staining and painting his crown molding, I decided to

laminate it. As it turns out, Kreutzfeldt decided to paint it (photo above).

Making the bending form—The first step was to make a form that matched the curvature of the convex wall. Finished with drywall, the wall defined a 90° arc having a radius of about 24½ in. Unfortunately, the curve was far from perfect, wandering out of round by up to ⅜ in. That forced me to make a template for the form.

To create the template, I bandsawed a 24½-in. radius curve in a sheet of ⅛-in. tempered Masonite. I then held this template against the curved wall 3¾ in. from the ceiling, which is where the bottom of the crown molding would contact the wall. After scribing the template with a pencil compass, I trimmed it with a jigsaw for a snug fit.

Molding in the round. **Made with basic shop tools, the laminated crown molding seen above wraps around a slightly out-of-round convex wall, butting at both ends into straight, factory-made crown. Photo by Rich Miller.**

Making the bending form itself was the easy part. Back at my shop, I traced the outline of the template onto three pieces of ¾-in. plywood, cut them out and nailed them together, placing ¾-in. plywood spacers between layers to produce the proper thickness. Finally, on the back edge of the form, I bandsawed a series of steps parallel to the front edge to give the clamps good purchase (photos next page).

Preparing the stock—The next step was to prepare thin strips of wood for lamination. The factory-made crown molding I wanted to match was made of white pine, but I selected clear sugar pine for my molding. Sugar pine has a uniform straight grain, is easily bent without splitting, and well, that's what I had on hand.

I produced the laminating stock by resawing ⅞-in. by 4-in. wide boards, which produced thin boards that were 3/16 in. thick. I used my bandsaw for ripping the boards into strips because its 1/16-in. saw kerf wastes less wood than my table-saw blade. Then I ran the strips through a 10-in. bench planer to remove irregu-

Glue up. The laminating form consisted of three layers of ¾-in. plywood separated by ¾-in. plywood spacers (photo above). A series of steps bandsawn into the back of the form provided solid footing for an assortment of clamps. Extra strips of pine (photo below) placed against the molding stock helped distribute the clamping pressure.

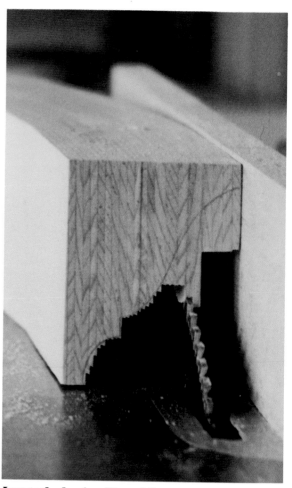

Low-tech shaping. The molding was contoured by making a series of cuts on a table saw to remove the waste up to the layout line. A mark on the rip fence indexed the center of the saw blade, indicating the optimal spot for the molding to contact the fence during the cutting operation.

larities and to reduce the strips to a uniform thickness of ⅛ in.

Next, I traced the cross section of the factory molding on graph paper. Examining this profile, I decided to laminate the trim out of three different widths of sugar-pine stock to simplify the removal of waste from the laminated blank.

With the strips cut to width, I dry-clamped them to the form to identify and eliminate any problems before glue up. I discovered that keeping the strips aligned would be difficult. My solution was to trace a slightly oversized cross section of the three-step assembly on two scraps of plywood, cut the patterns out and then slip the scraps over opposite ends of the assembly. These simple jigs helped to prevent the wood strips from sliding around during glue up.

Shake it up—For laminating jobs, I like to use urea-formaldehyde glue, which starts as a tan-colored powder that must be mixed with water (for more on builders' adhesives, see *FHB* #65, pp. 40-45). I used to employ a stick or a rubber spatula for mixing but sometimes ended up with lumps of powder that wouldn't dissolve. Then I discovered a better method: Put the powder in a plastic container, add the correct amount of water and then shake the container in a circular motion. Surprisingly, mixing in this way is faster than with a stick and produces a lump-free mixture every time.

To make glue up simpler, I taped the pine strips together edge-to-edge on my glue-up table. Then I spread the glue across the assembly using a paint roller. That done, I removed the masking tape, coated the masked areas with glue, tilted the strips upright and pressed them together.

Glue up took every clamp I had (including C-clamps, bar clamps and pipe clamps), and it wasn't a pretty sight (photos facing page). I installed the first clamp at the midpoint of the form and worked my way toward both ends, alternating clamps above and below the form. Extra strips of wood placed against the outer plies of sugar pine helped distribute the clamping pressure evenly. Excess was scraped off before the glue cured.

After letting the glue cure for 24 hours, I removed the blank from the form. Checking the concave side of the blank against the template, I saw that the molding was within ⅟₁₆ in. of a perfect fit. I decided not to shave it further just yet.

Sculpting on the table saw—Once the glue up was completed, I ran the curved blank, top-edge-down, through the thickness planer to remove slight irregularities from the bottom edge of the blank. Then I flipped the blank over and planed its top edge to size. Planing the curved blank was easy—I simply steered it through the planer to keep it perpendicular to the cutterhead.

With the sizing completed, I squared both ends of the blank and traced the outline of the factory crown molding on one end. Now all I had to do was remove everything that didn't look like crown molding.

Probably the easiest way to make crown molding is to cut it on a shaper. Many cabinet shops nearby had a shaper, but none had a cutter

that matched my molding. I could have ordered custom-made cutters, but that would have cost $300 to $400, difficult to justify for a one-off piece of trim.

I decided to shape the blank by making a series of table-saw cuts to remove most of the waste (photo facing page). This worked remarkably well. I clamped a 3¾-in. tall board to the fence to make it the same height as the molding, then marked the top of the fence to index the centerline of the saw blade. While making each cut, I held the blank against the fence at the index mark. Any deviation from this mark was insignificant, because it merely caused the saw blade to wander into the waste area, requiring nothing more than a second pass across the table saw to get it right.

My blade cut a ⅛-in. wide kerf, so I made the cuts by moving the fence toward the blade in ⅟₁₆-in. increments, raising the blade just enough each time to remove the maximum amount of stock without cutting across the layout line. In this fashion, I finished with a cross section very close to that of the factory crown.

Because the stock laid flat on the saw table, the cutting operation was accomplished safely and easily. I also kept my fingers far away from the sawblade at all times.

Scraping it smooth—All that remained was to smooth out the small, sharp steps on the molding blank. I figured this would be the easy part, but it turned out to be the most difficult.

First I tried sanding, but the sandpaper quickly became clogged with pine resin. I soon realized I'd have to scrape the pine smooth.

To make a scraper, I cut a 45° angle on the end of a scrap piece of the factory molding and traced its profile on the blade of an old taping knife. Then I cut the blade to the layout line using a bench grinder and raised a cutting edge by rubbing the blade with a hardened-steel punch.

Dragging this homemade scraper along the molding at a 45° angle produced satisfying re-

sults (photo below). Each pass left behind a small trail of fine dust instead of the curled shavings a perfectly tuned scraper would produce, but little by little the sharp steps began to disappear. Scraping the molding down to the layout line took two hours and a lot of elbow grease.

The scraping left the molding with some torn fibers and minor irregularities, so I decided to finish the job by sanding. I began by using 80-grit sandpaper to work out the irregularities, then worked my way up to finer-grit paper. I use Wetordry TRI-M-ITE sandpaper (3M Construction Markets Department, 3M Center, Bldg. 225-4S-08, St. Paul, Minn. 55144; 612-736-7761) because its backing doesn't tear while sanding. The sanding took about three hours and just about wore me out.

Installation—Earlier, when trimming the ends of the curved blank, I had left the ends 1½ in. long. Now I used the bandsaw to cut a 1½-in. long triangular stub tenon on both ends of the molding. These tenons would fit into the triangular voids behind the factory crown molding, aligning the joints while providing solid backing for the ends of the factory molding.

The final step was to fit and attach the curved crown to the wall. Using a belt sander, I relieved the concealed edges a bit so that the molding fit snugly against the wall and the ceiling. By now Kreutzfeldt had decided to paint the crown molding, but just a small amount of belt-sanding produced such a tight fit against the wall and the ceiling that no caulking was necessary. I applied construction adhesive along the back and top of the curved crown, then fastened the molding to the wall with screws run through the stubs.

As Kreutzfeldt installed the straight runs of crown molding, I was gratified to see that just a bit of sanding produced a satisfying match of curved to straight molding. ☐

John La Torre, Jr., is a carpenter in Tuolumne, Calif. Photos by author except where noted.

A homemade scraper. The author smoothed the sawcuts using a scraper made out of an old taping knife. The edge of the scraper was shaped with a bench grinder.

Making Classical Columns

Interior detailing with a router and lathe

by Joseph Beals

Classically detailed columns are commercially available in almost any size and configuration, but they're expensive, particularly when it comes to custom orders. And because the correct proportioning and form of classical design isn't common knowledge, local fabrication may not be a practical option. In addition, even a modestly sized column can easily exceed the capacity of most small-shop lathes.

To make the two load-bearing columns, which completed a room divider between our kitchen and living room (photo below), I devised a method of turning and fluting column shafts with a shop-built jig that uses a rail-guided router as the cutting tool. Each column has six separately made components (drawing facing page). The base plinth at the bottom and the abacus at the top are square sections. Except

for the router-turned shaft, the remaining parts were turned on a lathe.

A hybrid design—The columns I designed combine a clasically correct 24-flute Ionic shaft with an Attic base, a plain necking and a Roman Doric capital. This hybrid configuration is quite common; it captures the grace of Ionic columnation without the need for a hand-carved Ionic capital, a detail that can overpower the design, the budget, or both.

Classical columns are typically proportioned in modules, with one module equal to the bottom diameter of the column shaft. The module system is more a guideline than a set of inviolate proportions. Every text introduces variations, but there is very little agreement on the details. According to Giacomo Barozzi da Vig-

nola, whose *Comparison of Orders* (written in 1563) is often cited as the rulebook of classical design, an Ionic shaft should be seven modules high (not including the necking, which is typically one more module high), or in this case 49 in. To compensate for the relatively short hybrid capital, and to keep the length of the necking within proportional limits, I increased the shaft height to 52 in. The design of the capital is quite simple and substantially correct, but is detailed to echo the trim at the tops of the adjacent pilasters.

According to tradition, the bottom third of the shaft should be of a constant diameter, while the top two-thirds taper inward. This taper, or entasis, eliminates an optical illusion that makes a straight shaft appear narrowed in the middle. Because the columns are comparatively short, I modified the entasis ratio to avoid excessive taper, which might create an awkward appearance. Vignola gives $\frac{5}{6}$ of the module as the correct diameter for the top of a shaft, which in this case would have equalled about 5.83 in. I increased the diameter to eleven-twelfths of the module, or 6.42 in., which I rounded off to $6\frac{3}{8}$ in.

Coopering the shafts—Because the shafts posed the greatest challenge, I constructed them first. I never considered making solid shafts because they would require a lot of stock and be heavy, cumbersome and potentially unstable (solid columns have nowhere to expand but outward in response to increased humidity). My instinct was to attempt coopered construction, rather than the simpler but less elegant bricklaid alternative. Commercial columns are almost always coopered, but with glue-jointed staves tapered so that, once the shafts are turned, the thickness of the shaft walls remains constant from top to bottom. Because my columns would be relatively short, with a minimal taper, I deemed that degree of sophistication impractical.

I chose a straight-staved hexagon as a congenial compromise. By using $1\frac{3}{8}$-in. thick poplar that I had on hand, I could ensure sufficient wall thickness to accommodate a fluted, tapered shaft without compromising strength. I considered splining the staves, but the logistics of gluing 12 edges and 6 splines and assembling everything quickly and gracefully was a terrifying prospect. Having decided on a plain mitered joint, I ripped all the stave stock

The fluted columns, combined with raised-panel pedestals, pilasters and an ornamental header, frame the opening between the author's kitchen and living room. The hybrid columns blend an Ionic shaft with an Attic base, a plain necking and a Roman Doric capital.

Photo: Staff

slightly oversize, set the staves aside for a few days, then jointed them and ripped a 60° bevel on each side. Now I was ready for glueup.

Rounding up the staves—Gluing each shaft was surprisingly simple. I cut the staves to length, laid six of them across a pair of sawhorses, and quickly applied yellow carpenter's glue to the beveled edges. One by one, I stood them on the floor to assemble the hexagon— my left hand on the tops of the staves as I placed them together, my right hand reaching for the next piece and setting it in place.

When the hexagon was complete, I tightened a large stainless-steel hose clamp around the top to hold the parts together. Then I used a combination of hose clamps and a heavy Jorgensen band clamp (which applies tremendous circumferential pressure) to pull the staves together down the rest of their length. Working my way down the shaft, I drew the staves tight with the band clamp, secured a hose clamp beside it, and then moved the band clamp to the next gluing position. I repeated this operation seven times.

I glued the second shaft the next day, and again the operation went smoothly. Because clamp pressure is applied evenly across the six arrises formed by the 12 butting edges of the staves, the effect was to force the staves into perfect alignment. Because the arrises would be removed in turning the shafts, damage from the clamps was of no consequence. As it turned out, splining the staves would have complicated the preparation considerably and done nothing to improve the results.

A shop-built router jig—My bench-top router jig consisted of an "axle" suspended between two turning centers, and a pair of rails above the axle that supported and guided the router (photos 1 through 6, next page). The axle was a length of ½-in. galvanized iron pipe, threaded at one end. With an outside diameter of almost ⅞ in., the pipe was stiff enough to support a column shaft without flexing. The shafts were attached to the axle by hexagonal plywood plates that were let into either end of the shaft and screwed to circular plywood end-caps. These two-part wheels were drilled through their centers with a ⅞-in. bit and fitted with a simple wooden clamp that locked the shafts to the axle (photo 1). The turning centers consisted of a pair of plywood brackets screwed to the top of my workbench.

A few construction details were incorporated to make the jig reliable and convenient. The head end of the axle was fitted with a 12-in. dia. pulley aligned with the headstock pulleys of an old Delta lathe that sits near the end of the bench (photo 2). The pulley was belted to the smallest headstock pulley, and the lathe was set for its slowest speed. That produced about 200 rpm on the column shaft, which turned out to be the upper limit of the router's capacity for a light turning cut. To provide a smooth, relatively non-wearing bearing surface, I inserted pieces of copper flashing beneath the pipe axle where it passed through the

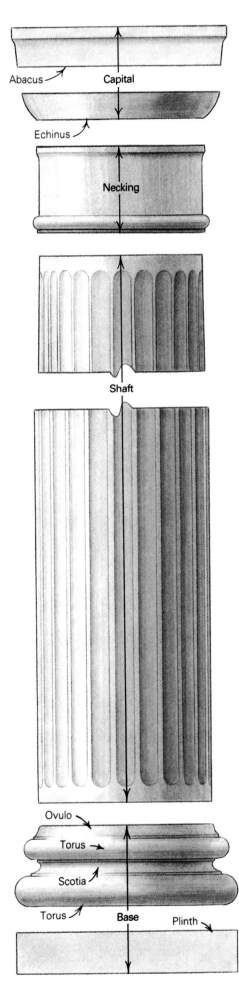

plywood turning centers. With a few drops of oil to lubricate the action, the shaft turned easily, and the holes didn't enlarge and become sloppy before the critical fluting operation.

Finally, a pair of 2x guide rails were sawn and planed to match the entasis, or taper of the column shaft (photo 4, p. 63). The inside edge was rabbeted to provide a bearing ledge for a plywood router base; the rabbet contained the router between the rails, but allowed it full axial movement. The rails were screwed to the turning centers high enough above the centerline of the column shafts so that the cutting bit would just clear the arrises, but not so high that the fluting bit would need to be extended excessively from the collet.

Turning the shafts—Shaft turning was simple but tedious. The router was fitted with a ½-in. carbide-tipped, hinge-mortising bit, with the initial cutting depth set to hog off as much material as possible without bogging down the router. I quickly discovered that it wasn't practical to make the roughing cuts with the shaft driven by the lathe; the router would chatter and dance as each arris swept beneath the bit, proving the machinist's axiom that an interrupted cut is the most difficult of all turning operations. Instead, I did the rough cutting with one hand guiding the router and the other turning the shaft, using whatever combination of motions was least tiresome at the moment. Wasting the bulk of stock across the arrises took some time, particularly at the smaller top end of the shaft where the cutting depth penetrated well into the flats between arrises.

Only when the shaft was quite round and very close to its final dimensions did I make a finish cut with the shaft driven by the lathe. This cut was accomplished by advancing the router slowly along the guide rails about five minutes to traverse the column. The result was close to perfect. Hand-sanding quickly removed the few traces of tool marks.

Fluting the shafts—The threaded end of the axle was fitted with a standard pipe cap through which I drilled and tapped a ⅜-in. fine-thread hole. That allowed me to bolt a 24-tooth circular-saw blade tight to the pipe cap, secured by a toothed lock washer under the bolt head to ensure a non-slip fit (photo 3). A simple lever and stud screwed to the plywood turning center engaged the gullets on the sawblade, serving as a simple but very accurate indexing head for cutting the column's 24 flutes. For safety's sake, the blade should *not* be mounted until just before the fluting operation.

In half the photographs and drawings I studied, Ionic flutes were of constant width, and the narrow fillets between them tapered toward the smaller diameter at the top of the shaft. In the other photos, the fillets were of constant width and the fluting was tapered. There is a subtle elegance to tapered flutes, but machining them would require changing the taper of the guide rails when changing over from turning to fluting. That was more trouble than I was willing to endure. A constant-width flute produces a tapered fillet, so I needed to determine the maximum flute width by defining the minimum acceptable fillet width at the top of the shaft.

By referring to more illustrations, drawings and my own instinct, I chose a strong $\frac{1}{8}$ in. as the minimum fillet width. That produced a flute width of about $\frac{11}{16}$ in. and a fillet width at the base of the shaft of a bit more than $\frac{1}{4}$ in., which was visually ideal. The router was fitted with a 1-in. carbide-tipped, core-box bit, which would cut the desired flute width without going too deep. The stops were tacked to the guide rails to hold the flutes about $\frac{1}{2}$ in. from the tops and bottoms of the shafts. I clamped the indexing lever into a sawblade gullet and set the router depth for the cut.

In contrast to the tedium of turning, milling the flutes was akin to magic. The first flute gave a hint of classic transformation. As the flutes began to march around the shaft, not only was the width perfect, but the shaft itself changed from a simple turning to a noble architectural feature. When the first shaft was finished, I took it down and put the second coopered hexagon into its place. Again I suffered the ordeal of turning, and then emerged from the noise and sawdust of a cobbled-together workshop device into the stratospheric realm of classic design. When it was over, the two shafts had taken on a life of their own.

Capitals and bases—I had enough stock left over from making the shaft staves to glue up two short sections for the one-piece column neckings. Each necking is turned to the same diameter as the top of the shaft, with an astragal at the base and a slight flare to a shoulder at the top. The first necking was turned in the conventional way on my lathe. It then served as a model for the second, and was duplicated with the aid of a caliper and a few pencil lines.

It remained to turn the 1-in. high echinus (the broad flare at the bottom of the capital that abuts the top of the necking) and the cylindrical portion of the base. In conventional practice these parts would be made out of solid sections with the long grain horizontal. But I was concerned with the problem of hiding alternating long grain and end grain under a finish coat of enamel. To avoid future regrets, I made the turnings so the long grain would be oriented vertically. This would produce a narrow band of end grain around the circumference of the torus, but at least it would be continuous.

I laminated seven lengths of 1⅜-in. thick poplar to make a cube almost 10 in. on a side. I crosscut the cube with a bandsaw to produce a short section for each echinus and a large section for each base. Because the long grain is discontinuous across the seven laminations, the sections should be relatively stable. They were screwed directly to the faceplate of the lathe, and turning was a straightforward matter of following the plan sketch.

The bottom of each column base (called the plinth) and the top of each capital (called the abacus) is a square section. Typically,

Jigging and assembly. The column shafts were turned and fluted with the use of a shop-built, benchtop router jig. The core of the jig was a ½-in. galvanized iron pipe that suspended the shafts between a pair of plywood turning centers. The hollow shafts were plugged at both ends with a combination plywood cap and wooden clamp that accommodated the axle and clamped the shafts to it (photo 1). For turning, the axle was fitted with a 12-in. pulley belted to an adjacent shop lathe (photo 2). At the opposite end, a 24-tooth circular-saw blade (mounted backwards for safety) served as an indexing head (photo 3).

these parts also would be made solid with the long grain horizontal, once again creating problems with dimensional stability. I turned to cabinetmaking methods to avoid trouble. Each plinth and abacus consists of a mitered frame made of 1⅝-in. high by 1½-in. wide poplar, with tongued diagonal corner blocks let into a groove ploughed around the inside perimeter of the frames (photo 5, below). The corner blocks reinforce the glued miter and fill the gap that would otherwise be exposed at the abutting turned sections. No end grain is exposed, and the narrow stock avoids the instability inherent in a solid block.

Using a shaper, I milled each abacus with a cove that ends at a shoulder to complete the top of the capital. The base plinths remain square, but were made slightly taller than I really needed to allow for trimming the columns to length later on.

Assembly—The six components of each column were assembled with hidden screws and hexagon keys. The plinth is screwed to the bottom of the turned section of the base through the corner blocks; the square abacus is screwed down through its corner blocks to the turned echinus. The remaining assembly has friction-fit hexagonal keys cut from ¾-in. plywood (photo 5). A key screwed to the top of the base engages the shaft, and a key screwed to the bottom of the echinus secures the necking. A pair of hexagonal keys screwed together align the necking with the top of the shaft (photo 6); the keys are offset by 30° to prevent them from falling into the shaft.

The keys center each component to its neighbor. They also ensure that a flute is perfectly aligned with the middle of the four square sides of the base plinth and the abacus. This is a traditional detail that seems trivial unless it's overlooked. A displacement will be hard to identify as a fault, but it gives the subtle and peculiar appearance of something not quite right.

The columns were primed, sanded, given a first finish coat and lightly sanded again before installation. At the site, I cut a measuring stick to fit at each of the two column locations. Each column was assembled on the floor and the base plinth trimmed until the column was a strong 1/16 in. taller than the measuring stick. I used a small hydraulic jack and a post to lift the header until each column tucked into place. When the jack was removed, the columns took enough load to hold everything solidly without the need for any further fastening.

The columns were given a final finish coat of oil-base enamel. The result gives a powerful, classic appearance to what before had been an attractive but incomplete design. A few visitors have noticed the columns at once, but most have stood back, puzzled at the dramatic but hidden change. "This looks great," one of them said. "What did you do—have everything repainted?" □

The router was fitted with a square plywood base and guided by a pair of rabbeted rails screwed to the plywood turning centers (photo 4). For dimensional stability and to hide the end grain, the base plinths (photo 5) are made of mitered frames reinforced with corner blocks. The abacus atop each capital is made the same way. To assemble each column, the author screwed the plinth and abacus to the adjacent turnings, then joined the rest of the components with friction-fit ¾-in. thick hexagonal plywood keys—double offset keys to join the neckings to the shafts (photo 6), and single keys to join the rest of the parts (photo 5).

Joseph Beals is a designer and builder who lives in Marshfield, Massachusetts. Photos by the author except where noted.

Retrofitting a Threshold

A three-piece threshold provides extra weather protection, especially in exposed locations

by Gary M. Katz

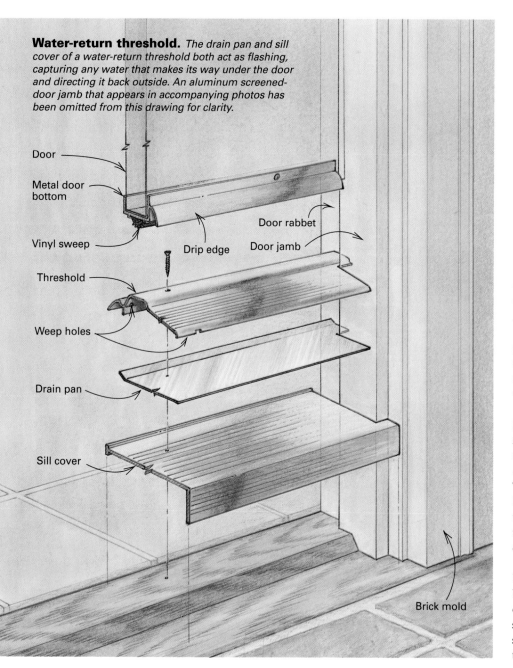

Water-return threshold. *The drain pan and sill cover of a water-return threshold both act as flashing, capturing any water that makes its way under the door and directing it back outside. An aluminum screened-door jamb that appears in accompanying photos has been omitted from this drawing for clarity.*

Door

Metal door bottom

Vinyl sweep

Drip edge

Door jamb

Door rabbet

Threshold

Weep holes

Drain pan

Sill cover

Brick mold

I used to install the ordinary type of metal thresholds available at hardware stores. Every time it rained, I'd worry. I'd worry about water sweeping in as the door swung open, water trickling in around the sides of the door or water entering through the screw holes. I'd worry about water warping a hardwood floor or staining a Persian rug.

Now I use ordinary thresholds only in protected openings. Experience has taught me that three-piece water-return thresholds are the safest bet. I often use thresholds made by Pemko Manufacturing (P. O. Box 3780, Ventura, Calif. 93006; 800-283-9988). A water-return threshold (drawing left) consists of a threshold, a drain pan and an interlocking sill cover. Although a water-return threshold is a little tough to install, the techniques I use make it simple enough, and the extra effort is worthwhile because it saves all of that worrying when it rains.

This example involves a door and frame already in place. While new, prehung-door units typically come with serviceable thresholds, the techniques discussed here could be used to add water-return thresholds to new doors.

Start with the sill cover—Sill covers are lifesavers. They are essentially the flashing for the threshold and cover the rough edge of a concrete slab or the exposed grain of a wood floor. They are also the perfect cure for elevation problems that can be created when, for example, a tile floor is laid right up to an original oak threshold and oak sill. This problem was the case in the door opening featured here.

I start by deciding how the sill cover should be notched around the jamb and exterior trim. Usually the sill cover butts against the door jamb or the brick mold (drawing left). On this job, a screened-door jamb had been added, and the sill cover had to remain behind that jamb so that the screened door would shut (photo 1).

Drawing: Dan Thornton

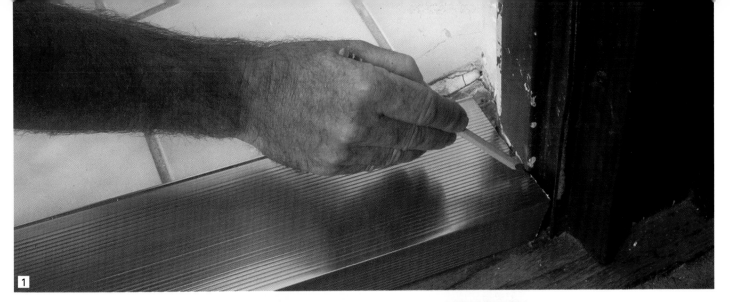

First, I cut the sill cover off square at the longest dimension needed, in this case from brick mold to brick mold. To cut the sill cover, I use a small circular saw equipped with a metal-cutting blade. (For more on cutting aluminum, see sidebar, p. 67). With the sill cover cut off square, I tip it into the opening and align it with the back of the screened-door jamb and with the rabbet for the main door. Using a pencil or a utility knife (scratch marks made by a knife are easy to see on most aluminum products), I scribe marks for the notch (photo 2). I repeat the process for the opposite side of the jamb. I'm using the jamb in place of a tape measure and square.

Slope the sill cover to drain—After cutting the notches, I set the sill cover in place and prepare to trim the front, or vertical, edge of the cover. On some openings this step isn't necessary. But if there's a concrete porch or wooden step just beneath the sill of the door, then the sill cover has to be scribed in. The cover must fit tight to the original sill, and it must have some slope so that water will drain outside, not inside.

I tip the sill cover and check the slope with a torpedo level; between 1/8 in. and 3/16 in. of pitch across the width of the sill cover is usually enough (photo 3). Using anything handy, I shim the sill cover in place. Then, on the inside of the opening, I use my square to measure the distance between the sill cover and the floor beneath. I spread my scribes accordingly and scribe a line across the front of the sill cover (photo 4). Sometimes I attach a clean piece of masking tape to the sill cover to make the line easier to see. I put on my goggles and earplugs and, holding the sill cover as far from my face as possible, I cut to the line with my circular saw.

Start with the longest dimension—I start fitting the threshold the same way I fit the sill cover, by measuring the widest dimension of the

Notch the sill cover around the door frames. First, the broad, flat sill cover is held in place to mark the location and depth of the notch (1 and 2) that will allow the sill cover to fit around the jambs of the exterior door and the screened door.

The sill cover slopes to the outside. The sill cover must be canted so that water can drain to the outside (3). Consequently, the front edge must be lowered by scribing it to the existing threshold, which is left in place (4). Masking tape makes the scribe line easier to see.

Scribe the threshold to the door frame. Rather than taking measurements and then transferring them to the threshold, the author holds the threshold itself against the jamb and carefully marks the locations of notches with either a pencil (5) or a knife (6).

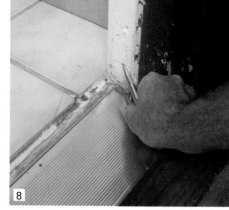

Door height is transferred from the jamb to the door itself. Once the sill cover and threshold have been notched and set in place, the height at which the bottom of the door will be cut off can be determined (7). The mark made on the jamb takes into account the thickness of the metal-door bottom (8). The author then scribes the bottom of the door (9) to the new sill cover. Masking tape on the bottom of the door makes the pencil line easier to see.

door opening, the rabbet for the main door. After making the first cut for overall length, I slide the threshold into the opening to mark the notch (photo 5).

Normally, the threshold aligns with the face of the door, but for this opening I wanted to pull the threshold inside the house ¼ in. so that it would cover the raw edge of the tile. I tip the threshold, hold it against the jamb and mark the notches (photo 6). I repeat these steps for the opposite end and cut the notches.

Once the threshold is cut, I temporarily set it in place on top of the sill cover. I mark the spot where the front edge of the threshold rests on the sill cover. This mark will determine the location of the drain pan, which is installed between the threshold and the sill cover. It's important to locate the drain pan carefully so that it catches water seeping through weep holes in the threshold but at the same time remains hidden from view.

I set the drain pan in position just behind the mark on the sill cover, then scribe a line for the notch I need to make around the jamb. The thin drain pan is easy to cut with tin snips.

With the drain pan cut and in place, I set the threshold on top of it and drill pilot holes for the screws that hold the assembly to the floor. If I'm working on a concrete slab, I run my masonry bit through the threshold, drain pan and sill cover, down into the concrete. That's the surest method I know of getting concrete anchors in the right spots.

Cutting off the door—With the threshold and the sill cover in position, I'm ready to determine how much to cut off the bottom of the door. In order to get a weather-tight seal, this type of threshold requires a separate U-shaped metal door bottom with a vinyl sweep and drip edge (drawing p. 64); in this case I used one made by Pemko Manufacturing. In some installations, there is enough room for the metal door bottom between the bottom of the door and the new threshold. In this case, though, the door is too close to the threshold and has to be trimmed slightly. First, I measure up from the top of the threshold ½ in. and make marks on both jamb legs (photo 7). Then I remove the threshold and drain pan, but I leave the sill cover. The cover provides a smooth surface for me to run my scribes along.

I spread my scribes from the sill cover up to the line I've made on the jamb (photo 8), then shut the door and scribe a line across the bottom of the door (photo 9). Again, masking tape makes it easier to see the line.

I use different methods for cutting off doors. On veneered doors I sometimes use a "shooting

stick" straightedge made of thin plywood (photo 10). The shooting stick allows me to cut doors quickly without worrying about tearout. If I don't have my shooting stick with me, I use a metal straightedge and a knife to score a line on the door before I cut it. Either way, it's my circular saw that does the hard work.

Seal the threshold with plenty of silicone— Before I install anything, I sweep out all of the dust and dirt, especially under the sill cover. I run a bead of silicone under the sill cover to help secure it, though the screws that pass through the threshold really do that job. I press the sill cover into the silicone and then run another bead of silicone on top of the sill cover and beneath the drain pan. I take care to keep the silicone away from the front edge of the drain pan so that it doesn't squeeze out. Then I press the drain pan down into the silicone and apply more silicone at the joint of the jamb and drain pan. I also squirt silicone into the screw holes. Finally, I set the threshold and screw it down snug.

It's better to notch the metal door bottom around the weatherstripping on the door jamb, so I install the weatherstripping first if there isn't any already. The door bottom has to be cut to fit the overall width of the door and then notched to fit around the weatherstripping on the door jambs (photos 11, 12).

Not too tight, not too loose—After cutting the door bottom, I slip it on again and swing the door shut to make sure everything fits. The door bottom shouldn't be too long and squeeze or rub against the weatherstripping, but it should come close. I press the door bottom down against the threshold, but not too hard. The vinyl sweep needs to contact the threshold, but shouldn't be forced against it. Otherwise, the sweep will compress over time, and the seal will be lost.

I drive one screw at each end of the door bottom to hold it in place. Then I check the action of the door to be sure that it's sealing but that it's not rubbing too hard. Then I drive in the rest of the screws.

All that's left then is to apply silicone to the threshold and sill cover at the joint of the jamb, and maybe a little caulking between the door bottom and the door to seal out moisture. Before I leave, I check the swing of the door one more time. When it's right, the door closes just like a refrigerator, and with that whoosh of air, I'm gone. ☐

Gary M. Katz is a carpenter/contractor and writer in Encino, California. Photos by the author.

A site-built tool speeds the cut. The author cuts doors with a "shooting stick," a plywood straightedge that has a fence against which he registers the table of his circular saw (10).

First the weatherstripping, then the door bottom. Weatherstripping is installed on the side jambs first. The drip cap on the door bottom then is scribed and notched to fit around the weatherstripping (11 and 12).

Take care when cutting aluminum

Cutting aluminum thresholds is not my idea of fun, so I like to do it as quickly and as safely as possible. Many people use hacksaws. Some professional weatherstrippers use portable table saws with aluminum-cutting blades.

I use a 4⅜-in. Makita model 4200N trim saw (Makita U. S. A., 14930 Northam St., La Mirada, Calif. 90638; 800-462-5482) that cuts at a fast 11,000 rpm and a fine-toothed, combination metal blade, also by Makita (model 792334-2). I've used carbide-tipped blades to cut aluminum, but they are expensive, especially when the teeth begin to break off. (Makita no longer makes the 4200N, although some distributors still have a few of these saws. It was replaced with the model 5005 trim saw, which has a larger and slower-turning 5½-in. blade.)

Wearing eye protection is important when cutting aluminum, not only to guard against bits of flying metal but also because teeth on the combination blade can chip. In my work box I carry plastic goggles wrapped in a sock to prevent scratches. Years ago a bungee-cord accident took most of the vision in my left eye, so I'm careful with my right one. I also wear earplugs.—G. M. K.

More Than One Way to Case a Window

You can vary the look with simple combinations of flat boards and stock moldings

by Joseph Beals III

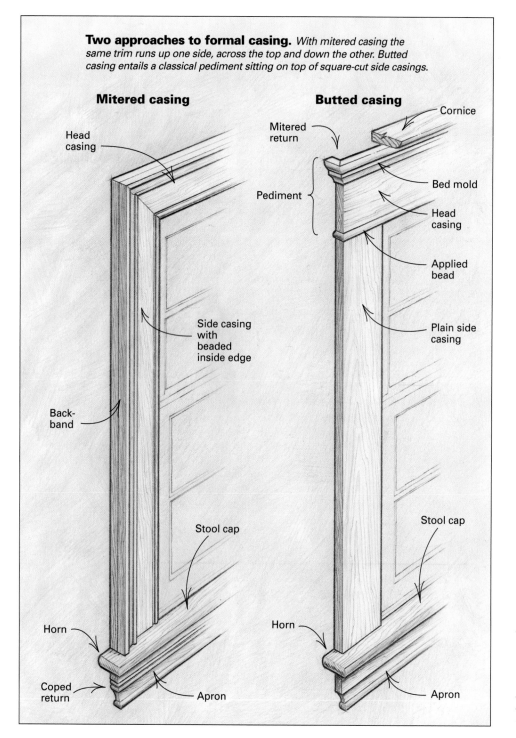

Two approaches to formal casing. *With mitered casing the same trim runs up one side, across the top and down the other. Butted casing entails a classical pediment sitting on top of square-cut side casings.*

Mitered casing

- Head casing
- Mitered return
- Side casing with beaded inside edge
- Back-band
- Stool cap
- Horn
- Coped return
- Apron

Butted casing

- Cornice
- Bed mold
- Head casing
- Pediment
- Applied bead
- Plain side casing
- Stool cap
- Horn
- Apron

Stock trim, such as 1x4 pine or clamshell casing, will always be useful in routine, contemporary construction. But after years of installing these lifeless, standard-issue casings, frustration drove me to cross the frontier. Using as a reference the casings I'd seen in so many period New England houses, I built a simple pediment head casing to surmount square-edge, square-cut side casings. I will not forget that first look back at the result: An ordinary window suddenly had character, grace and purpose; and a client new to classical trim was particularly pleased with a window worth looking at, not just through.

Mitered casings get moldings; butted casings get pediments—In this article, I'll discuss two approaches to formal casing design (drawing left): mitered casings as a stylistic option and butted casings, which employ a pediment head, or architrave, as a classical architectural option. In general, mitered casings are developed by adding layers of moldings to the perimeter of a mitered flat casing. With butted casings, the pediment head sits on the square-cut tops of plain or molded side casings, and architectural detail is developed on the pediment itself.

In a mitered casing the simplest alternative is the use of two or more layers of molding. A bead cut on the inside edge of a casing and a backband applied around the outside perimeter of a flat casing will give a strong, three-dimensional appearance (top right photo, facing page). A thin molding interposed between the flat casing and the backband adds another element of detail and shadowline to the profile (bottom right photo, facing page).

A formal alternative to mitered casings is the use of a pediment, or architrave, as a head casing with side casings butting into it. The pediment represents an entablature, the lower portion of a classical roof, and its elements are derived from ancient Greek and Roman temples. This style of window trim is commonplace in period architecture, especially in Federal and Greek Revival houses of the 19th century.

All window casings start with the stool cap—Any style of window trim must begin with the stool cap, the piece that finishes the inside

Drawings: Dan Thornton

Six variations on a theme

We asked designer/builder Joseph Beals to show us how to create various styles of window trim without resorting to custom moldings made with a shaper. On a test wall in his shop, Beals mocked up six casing designs, using mostly dimension lumber and stock moldings from the local lumberyard. He cheated occasionally and used his shaper, but where he did, a router would work just as well.

The stool must stay level and perpendicular. The author screws blocks of scrap wood to the wall to hold the stool temporarily.

edge of the windowsill. You can buy stock stool cap at a lumberyard, but you'll spend enough time fitting it to the window that you might as well make your own.

I like the stool to stand proud of the casings about an inch. If you're thinking about a two-piece or three-piece mitered casing, you'll have to figure out the combined thicknesses of the moldings before you cut the stool.

I cut the stool stock long, and I machine a bull-nose, or half-round, on the outside edge. I scribe and cut the two horns (horns are the ends of the stool that overlap the drywall on each side), check the fit, then cut the ends to allow about a 1-in. overhang past the outside edges of the side casings. The bullnose returns can be shaped by machine, but I prefer to remove the bulk of the waste with a block plane and finish the job with a sharp file or a piece of cloth-backed sandpaper.

Given the wide variety of window conditions, fastening the stool can be a problem. On the Andersen window used in these photos, the stool is toenailed from the top into the sill. The horns can be face-nailed into the framing studs if nec-

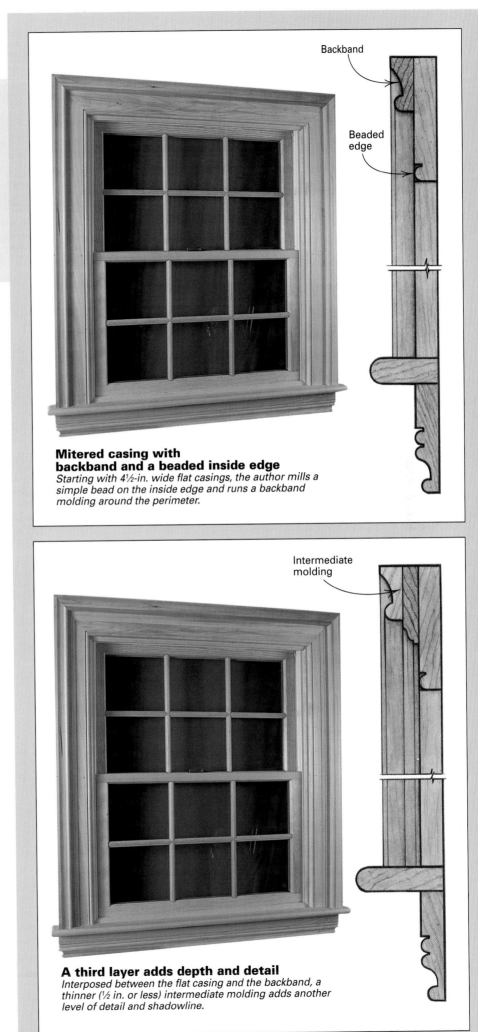

Mitered casing with backband and a beaded inside edge
Starting with 4½-in. wide flat casings, the author mills a simple bead on the inside edge and runs a backband molding around the perimeter.

A third layer adds depth and detail
Interposed between the flat casing and the backband, a thinner (½ in. or less) intermediate molding adds another level of detail and shadowline.

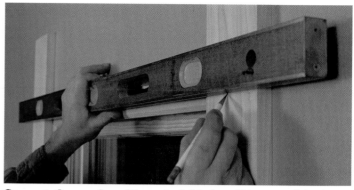

Mark the side casings with a gauge block. The author uses a scrap of wood, marked with the amount of offset between the window jamb and trim, to mark the length of the side casings.

Biscuits register side casings to stool. Common practice is to nail up through the stool into side casings. But biscuits hold better than end-grain nails and won't come out through the exposed surface of the side casing.

essary. Glue or caulk between the stool and the sill can help hold everything together.

To keep the stool square to the window, I screw 2x4 blocks to the rough framing right below the stool (photo left, p. 69). Nothing is worse than getting a window trimmed and realizing that during the course of your work, you've pushed the stool out of square with the wall. The blocks hold the stool square until I add the apron at the end of the job.

Cut the side casings long, fit them to the stool and mark their length—For any casing design, the two side casings are cut long and fitted first to the stool. I cut the casing bottoms square, then hold a casing in position against the window jamb. I judge the reveal by eye (the reveal is the exposed portion of the jamb between the casing and the jamb's inside edge), but if you prefer reference marks, make them with a gauge block of some kind.

In theory, if the bottoms of the side casings are cut square and if the stool is perpendicular to the sides of the window, the casings should fit tight to the stool, but theory doesn't always work. I close any gaps by dressing the casing bottom with a sharp block plane. After both side casings have been fitted to the stool, I tack them in position with a few 4d finish nails.

To mark for the top cuts, I use a gauge block as shown in the photograph (above left). For mitered casings, the mark I make indicates the bottom of the miter. I draw a diagonal reference

mark on the casing to remind me of the angle of the cut. This procedure may seem foolish, but it is remarkably easy to forget how things go together in the short walk to the miter saw.

When I mark the two sides of a butted casing, I make tick marks on the inside edges with the gauge block and then connect the two marks with a straightedge (photo above right) and a sharp pencil. In theory, the top cuts should be square to the casings, but this theory may not be true in practice, especially in the case of an old window that you're retrimming. Before nailing the side casings, I cut biscuit slots in the bottom ends and mating slots on the top of the two stool horns (photo bottom right).

Nail the side casings home before you fit the heads—Some people prefer to fit the head casings while the two side casings are only tacked in position on the theory that it's easier to make the joints perfect if both pieces are adjustable. Then, after everything is fitted, you nail the pieces home in one marathon effort. I think this process invites problems because the side casings can shift as they are pulled against the wall when the nails are driven home. Even setting a nail after it has been hammered in can cause the casing to shift, and the joint you thought was perfect opens up again. Adjustments to bad joints are awkward or impossible.

After I fit a side casing to the stool, mark it, cut it to length and cut the biscuit slot, I nail it tight and set all the nails. The side casings now are

finished; they have been locked in place so that they can't move.

Other people prefer to install mitered casings by working around the window: up one side, across the top and down the other side. I prefer my method because the two sides are done quickly and easily, and any adjustments are made in the head casing. Only if the window is badly racked would this be impractical, in which event it might be better to fix the window condition first.

Fitting a mitered head casing—I like to cut a bead on the inside edge of all of my mitered window casings. A shaper, router or cutter head-equipped table saw cuts beads equally well. The molded bead mimics the applied sash stop found on a lot of old double-hung windows.

For a mitered head casing, I make the 45° cut on one end and then mark the other end. As the photograph (left photo, facing page) shows, the casing is held upside down with its mitered end perfectly registered on the side casing, and the other end is marked with a utility knife.

I cut the second miter, saving the knife line. I drop the head casing in place, taking care to keep it parallel with the window head jamb. Any deficiencies in fit are apparent, and these flaws are dressed out with a sharp block plane.

I use biscuits to reinforce the mitered joints. The biscuit slots in the side-casing miters can easily be made in place if they weren't done earlier, and the head-casing miters can be slotted on any

For mitered casings nail home the two sides and then fit the head. The author makes a 45° cut on one end of the head casing, holds it upside down on the side casings and marks for the other cut with a utility knife.

flat surface. If all is well, I glue the slots, slide the biscuits in and tap the head casing into place.

More layers add depth and detail to mitered casings—As the photographs of finished casing styles show (see p. 69), mitered casings can be built up in layers to give a handsome appearance. The tight, solid base offered by the flat casings makes the application of additional components easy.

There are plenty of backbands available at lumberyards, or you can make your own with a table saw and router. I install the backband in a sequence similar to the flat casings: fitting and nailing both sides and then fitting and nailing the head piece.

A three-part casing with an intermediate molding between the backband and flat casing adds another level of detail to the casing. I make the intermediate molding from ½-in. or thinner stock. Making this piece from thicker stock can result in a clunky, heavy casing. A thinner intermediate molding has a more delicate look.

On these multilayered casings, the outside edges may need dressing to present a single, flat surface. I do this with a sharp bench plane (photo left, p. 72), working as close as I can to the stool horns, and I clean up the last few inches with a sharp chisel.

Pediments dress up butted casings—One of the beauties of adding a pediment to a butted casing—aside from the aesthetic ones—is that all

Simple butted casings
Plain 4½-in. wide side casings are topped with a 4½-in. wide head casing as a pediment. The only embellishment comes in the fact that the pediment is made from thicker stock (5/4) and overhangs the side casings in front and on the ends.

5/4 head casing

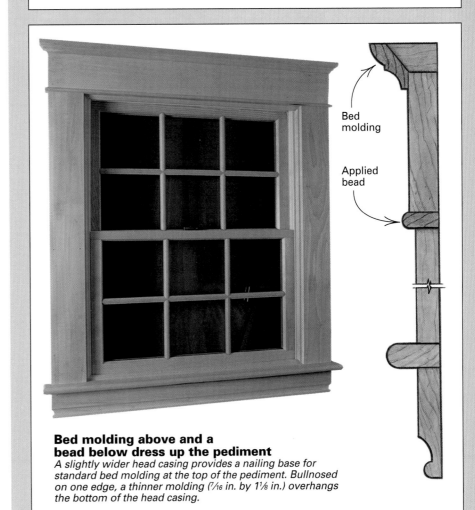

Bed molding above and a bead below dress up the pediment
A slightly wider head casing provides a nailing base for standard bed molding at the top of the pediment. Bullnosed on one edge, a thinner molding (⁷⁄₁₆ in. by 1⅛ in.) overhangs the bottom of the head casing.

Bed molding

Applied bead

Run mitered returns long; trim them with a handsaw. The author uses one hand to hold the handsaw while he makes the cut.

Plane the edges even. It's possible that the outside edges of all of the built-up moldings of a mitered casing won't be perfectly flush after they're nailed on. A couple of passes with a bench plane makes a flat surface. The last few inches at the bottom are cleaned up with a sharp chisel.

Nail the head moldings on a flat surface. To align the pediment's components, the author nails the pieces together on a flat surface.

of the pediments can be made on a bench and added as a complete unit to the tops of the two side casings.

A basic pediment is built from 5/4 stock, with an applied bead across the bottom and a bed molding across the top (bottom photo, p. 71). The space between the top of the bead and the bottom of the bed molding should be at least equal to the width of the side casings. You can make the space wider for a bolder appearance, but beware of overdoing it. Make a trial pediment if there is doubt about the aesthetic effect.

To provide a base for nailing on the bed molding, I make the height of the head stock at least ½ in. taller than the width of the side casings. The length of the head stock is equal to the distance between the outside edges of the side casings.

I draw a pencil line along the length of the head stock to indicate the bottom edge of the bed molding. With a square, I mark the bottom edge of each bed-molding return around the corner at each end of the head. This step is important because it's easy to cock a short return, and even a small misalignment will be brutally obvious once the pediment is installed.

The long piece of bed molding is mitered at each end. For convenience and safety, I cut the mitered returns long. I check the miters for a tight fit, then fasten the returns with glue and a few brads. To avoid nail holes, you can use masking tape to secure the returns while the glue dries. But this task is a slippery, three-handed job, and it

won't make much difference in the final result. I use a handsaw to trim each return flush with the backside of the head (photo above right).

To make the bead, I resaw a piece of stock $\frac{7}{16}$ in. thick by $1\frac{1}{8}$ in. wide, but these dimensions are not critical. I cut a bullnose along one edge with my shaper (also an easy router job). The bead overhangs the casings by about $\frac{3}{8}$ in. at each end. I shape the returns with a block plane, then finish the radius with a sharp file or a strip of cloth-backed sandpaper. The bead is applied to the head stock with glue and 4d finish nails. You can do the application free hand, but working on a flat surface like a table saw makes it easy to keep the back of the bead flush with the back of the head stock (photo bottom right).

Before installing the pediment, I cut biscuit slots on each side of the bottom, aligned with the centerlines of the two side casings. The slots in the side casings can be made before the casings are installed, but if you have enough ceiling height, it's just as easy to make them in place. Be aware of where you nail the bead onto the head stock so as not to put nails where they will interfere with cutting biscuit slots.

Pediments can range from simple to complex—As the photographs of casing styles show (see facing page), the pediment can be varied to increase the level of architectural detail. The simplest pediment is a plain piece of 5/4 stock, cut to a length that overhangs the side casings at

each end, as the bead does on the pediment described above (photo top right, p. 71). Adding a bed or cornice molding to the 5/4 stock begins to echo the lines of a true entablature.

The next evolution is a 6/4 head with a bed molding, surmounting a square-cut 5/4 fascia or lower molding (photo top right, facing page). Each element overhangs the face and ends of the element below, in the same style as the classical Greek and Roman entablature.

Adding a cornice molding above the bed molding in any pediment completes the basic elements for the full entablature (bottom photo, facing page). For interior cornices, I use 4/4 or 5/4 stock, and I machine the curve on my shaper with a knife I ground for the purpose. I shape the returns by hand, as described above for stool and bead returns, because machine work on short, end-grain sections is usually awkward.

If you choose to include a cornice, you should increase the height of the head stock by the full height of the bed or cornice molding and install the cornice molding first. Remember to make the cornice long enough to incorporate the bed molding returns. Install the bed molding tight to the cornice, and take care to seat it properly. It's easy to cock the long molding in rotation, which will make fitting the mitered returns frustrating.

Aprons complete the casings—An apron is nailed to the wall below the stool and is the lowest component of a window casing. On a lot of

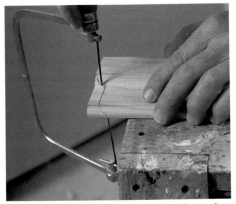

Cope the apron returns. Rather than miter returns, the author draws an outline on the molding profile and cuts it with a coping saw.

Different aprons fit different casing styles. The author adds the apron after the rest of the window is cased, experimenting with the apron design to find one he thinks works best with a particular window-casing style.

windows, a piece of casing stock serves as the apron, with its ends cut square, angled or returned, according to the whim of the carpenter or designer. There is no equivalent to the apron in classical architecture, which is why aprons in neoclassical window trim exhibit such a variety of profiles.

The apron has a specific aesthetic function, and there are several profiles that I use (bottom photo above). The apron visually returns the window to the plane of the wall below. About 3½ in. of height works well; if the apron stock is wider, it will look boxy and unbalanced, but narrower sections can be successful. I mill the profiles on the shaper and table saw, but a router will also handle this job easily. The simplest pattern echoes the backband that is used on the mitered casings. I make this piece out of solid stock, but a cove cut with an applied bead on the bottom is a simple alternative.

The apron returns are coped rather than mitered. I sketch the profile free hand, but you might prefer to use a paper or flexible plastic template. I remove the waste with a coping saw (top photo above), and I dress the return with a sharp file. This work goes quickly in contrast to mitered, glued returns, which are awkward and generally are not worth the effort, even on bright finished trim. □

Joseph Beals III is a designer and builder in Marshfield Hills, Mass. Photos by Jefferson Kolle.

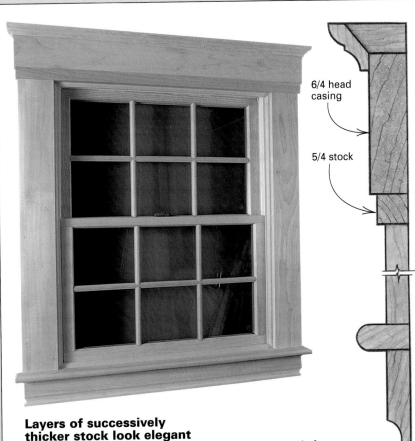

6/4 head casing

5/4 stock

Layers of successively thicker stock look elegant
Here, the ¾-in. side casings support a pediment composed of 5/4 square stock and 6/4 head casing, again capped with a bed molding. Each element overhangs the face and end of the element below.

Cornice

A cornice is the crowning touch
Adding a simple cornice piece above the bed molding and head casing completes the basic elements for a formal entablature.

Making Curved Casing

Strip-laminating arches to match straight casing profiles

by Jonathan F. Shafer

A few years ago I was asked to step in and complete the finish work on an 11,500-sq. ft. Tudor home that had taken 18 months to get through drywall. Completing the trim took an additional 12 months and posed many challenges, such as hanging 8-ft. high doors, building four stairways and running thousands of feet of wide casings and base. The house also had many arch-top windows (photo below) and doorways of various heights and widths. The curved window and door casings had to match the existing straight casing, so I decided to produce the curved casings on site with the help of a talented crew of finish carpenters.

My approach to this challenge was to strip-laminate the arched casings. By alternating strips from two pieces of straight, even-grained casing, we reproduced the casing profile. We ripped the strips from straight casing and then bent them around a form for each window and door. We also laminated extension jambs for each window, using the same bending forms.

Making the patterns—Our first step was to make patterns of all the arched windows and doors. How we produced the patterns varied depending on the particular application—some methods were as simple as tracing on kraft paper (available in long rolls) against the window frame, while others were as involved as mathematically computing arcs and multiple radius points.

One method we used on some of the more complex windows required a thin, flexible ripping of even-grained wood long enough to follow the arch along a window frame. This strip was clamped or held by helpers against the inside of the frame. We maintained the arch shape by tacking crosspieces to the bowed strip. The more crosspieces we used, the better the shape was held after the clamps were removed. We then transferred the shape of the arch to kraft paper.

With another method, we tacked plywood against the window, using a piece wide enough

to contain the unknown radius points. We then used a beam compass to find the radius points on the plywood by trial and error. Again, the arch was then transferred to kraft paper. I came on the project too late to have done it, but in the future I would make tracings of each window frame prior to installation.

Finally, we cut out each pattern and checked it against the corresponding window, making necessary adjustments. The patterns also had to be extended on both ends to allow extra casing length for trimming later. We labeled the patterns for window location and wood species.

Building the bending forms—When the patterns were ready, we built a bending form for each one. We constructed them from 2x stock cut into arcs on a bandsaw (for the design of a bending form for curved jambs, see the sidebar on p. 77). With roundtop casings, the 2x arcs were made using a simple circle-cutting

The curved casing for these windows was fabricated on site, using two pieces of straight stock cut into narrow strips and laminated around a form.

jig fixed to the bandsaw table (drawing below). We extended the table with a piece of ¼-in. plywood and ran a screw through it to create a pivot point. The 2x stock was then pivoted around the pivot point on a ¼-in. plywood carriage.

To cut the more gradual arcs of the bigger windows, we used a 1x3 to extend the pivot point of the circle-cutting jig across the shop (photo below). The 2x arcs were screwed to a plywood base or to the subfloor, depending on how big they were.

Ripping strips—Once the bending forms were completed, the strip-cutting operation was next. The basic principle here is that you're taking a piece of straight casing with the molded profile you want and ripping it into narrow strips that you can bend around a form and glue back together. But if you were to do this by ripping a single piece of casing, the resulting molding would be narrower than the original because of the material lost to the saw kerf. Therefore, you have to make alternate cuts on two pieces of straight casing.

To ensure that the laminating strips were cut to a uniform width, we used thin pieces of pine as spacers resting against a preset table-saw fence. This enabled us to cut the casing incrementally without changing the position of the saw fence.

In our case, the saw blade, and hence the laminating strip, was roughly ⅛-in. wide. The spacers were cut so that each was twice the width of the table-saw blade. We cut our spacers the same length as the short auxiliary fence on my table saw. To keep them from slipping with the casing as it was being cut we simply tacked a brad to the underside of each spacer, which hooked over the front edge of the saw table (drawing next page).

After the spacers were completed, we adjusted both pieces of casing (ripped a little off them) so that the finished width was an even number multiple of the spacers. Our casing had a rabbeted back band around the outside edge, so we were able to reduce slightly the outside edge of the casing without changing the profile. (The side casings were also adjusted in width to make them equal to the arched head piece.)

Next, we glued and clamped the back band to the casing. We also filled in the plowed relief on the back of the casing by gluing in thin material and jointing it flush. This was necessary so that each strip would be cut square to the others.

We set the table-saw fence to equal the total adjusted casing width (casing plus back band) minus the width of one spacer. Finally, we equipped our table saw with a riving knife (or splitter) mounted behind the blade. This protected the thin strips from damage as they came off of the saw.

In order to produce alternating strips from two pieces of straight casings, we ripped the first piece of casing on the saw with the outside edge against the fence (drawing next page). Next, the second piece of casing was ripped with the inside edge against the fence, and so on.

On the first piece of casing that we did, I made the mistake of dropping the laminating strips into a pile, thinking I could easily sort them out later. Wrong. It took me over an hour to put the pieces in order before gluing them up. For subsequent casing, I built a rack to hold the strips in order (top photo, next page). The rack was simply a pair of 1x4s with saw kerfs in them, nailed to a short bench.

Glue up—Before we could start gluing, the bending forms had to be adjusted to allow for the jamb reveal—the difference between the inside edge of the jamb and the inside edge of the casing. We tacked ¼-in. thick spacing shims (the width of the reveal) against the forms. We also covered the forms with waxed paper to prevent the casing from adhering to them during glue up.

We had plenty of pre-adjusted clamps on hand to do the job—everything from bar clamps to wedges against wood blocks screwed to the floor. Our clamping cauls were bandsawn to match the outside radius of the casing. We dry-fit the strips around the form so that we could work out a clamping strategy (bottom photo, next page). During glue up, we quickly and evenly brushed yellow glue on each piece. Because the casing was relatively wide and the set-up time relatively short, we glued and clamped the strips in three stages and let the glue dry overnight before proceeding to the

next stage. Once the complete casing was dry, we scraped and sanded the casing profile, removing glue squeeze-out and any irregularities in the profile.

Making extension jambs—We made the extension jambs for the windows with the same bending form used for the casing—all we had to do was remove the ¼-in. shims. The reveal was ¼ in. so we made the extension jambs ⅝ in. thick, allowing sufficient material to secure the casing. We produced strips roughly ⅛ in. thick to reduce the chance of springback and wide enough to fill the space between the window frame and the edge of the drywall, plus ¼ in. extra for ripping and jointing to the finished width after the glue had dried.

Ripping and jointing was a two-man operation. One man fed the piece into the table saw or jointer, and the other helped support the piece as it went into and came out of each machine. Cutting and fitting the extension jambs to proper length was a trial-and-error process, the error always being on the long side until the jambs fit. Next they were glued and nailed to the window frames through predrilled holes.

Fitting the casings—The ease of fitting the casings to the windows was directly related to the care with which the pattern had been made. If the pattern was true to the window form, the casing was relatively true to the window. Because our laminating strips were about

For round-top windows with small radii, Shafer cut sections of the bending forms on the simple jig shown in the drawing. For bigger windows, he attached a length of 1x3 to the plywood carriage (above) and extended it across the room to a center point on top of a workbench.

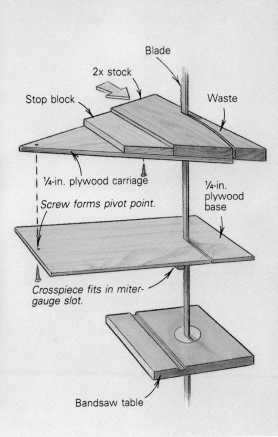

Bandsaw circle-cutting jig

Blade

2x stock

Stop block

Waste

¼-in. plywood carriage

Screw forms pivot point.

¼-in. plywood base

Crosspiece fits in miter-gauge slot.

Bandsaw table

⅛ in. wide, the springback was negligible. We were using relatively wide casings, so springing the casing to match the window—anywhere the pattern was not true—was very difficult, if not impossible. We had to live with compromises in a few places.

Just as with the extension jambs, fitting the casing was a trial-and-error process. It was relatively simple on the windows with one-piece casings that were butted directly to the window stools. Likewise, using plinth blocks would have simplified fitting the casing on the bigger windows and doors. But we decided to miter the corners between the arched casing and the side casings.

We calculated the miter by tracing the head casing and side casings right on the drywall, then connecting the points where their inside edges and outside edges intersected. The head casing was cut to match this line, and the corresponding angle was then cut on the side pieces. After the miter was judged to be tight, the bottoms of the side pieces were marked and cut square to rest on the window stool, or mitered if the window was picture-framed. □

—————————

Jonathan F. Shafer was a carpenter in Dublin, Ohio, when he worked on this project. He has since relocated to Bellingham, Washington. Photos by Kevin Ireton.

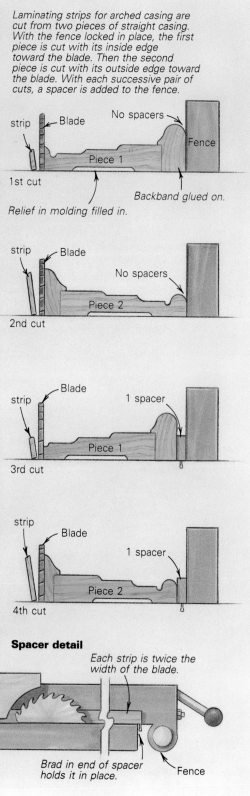

Cutting sequence

Laminating strips for arched casing are cut from two pieces of straight casing. With the fence locked in place, the first piece is cut with its inside edge toward the blade. Then the second piece is cut with its outside edge toward the blade. With each successive pair of cuts, a spacer is added to the fence.

1st cut

strip — Blade — No spacers

Fence

Piece 1

Relief in molding filled in.

Backband glued on.

2nd cut

strip — Blade

No spacers

Piece 2

3rd cut

strip — Blade — 1 spacer

Piece 1

4th cut

strip — Blade — 1 spacer

Piece 2

Spacer detail

Each strip is twice the width of the blade.

Brad in end of spacer holds it in place. — Fence

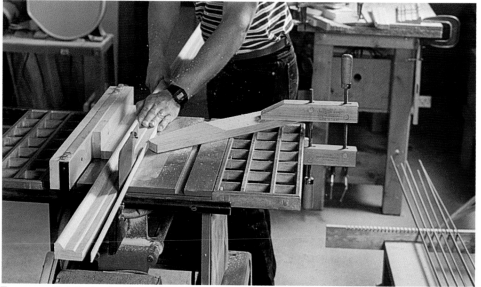

To ensure that the laminating strips were cut to a uniform width, Shafer kept the table-saw fence at a fixed distance from the blade and used spacer strips to move the straight stock incrementally closer to the blade. The rack beside the saw holds the laminating strips in their proper order for gluing.

The bending form is simply 2x stock cut into curved sections and screwed to a piece of plywood. Here the laminated strips have been clamped up without glue to work out the clamping strategy and eliminate some of the usual glue-up trauma.

Reusable jig for curved jambs

With the trend toward New Classicism in architecture has come a renewed popularity of arch-top windows and, to a lesser extent, doors. Making the jambs for these doors and windows presents only minor challenges. The millwork is straightforward, a matter of laminating a stack of plies in a curved jig. The problem is cost. One of the axioms in the millworking business is that no matter how many stock jigs you have cluttering up the back room, the next customer will want a window of a slightly different size. Building a custom jig for each order soon prices the work out of the market.

Over the last few years, David Marsaudon and the other folks at San Juan Wood Design have developed and refined a jig for making arch-top jambs that employs reusable parts. With this jig, they can lay up fair, smooth arcs of virtually any radius, and do so at a cost that gives them a competitive edge in the custom market. Their jig is made with some scrap plywood, strips of ¼-in. hardboard, a sheet of particleboard and a shop full of clamps. I visited San Juan's shop in Friday Harbor, Washington, and got a close look at how the jig works.

Building the jig—The jig, which somewhat resembles a dinosaur skeleton, is made of three parts: a semicircular particleboard panel, a set of clamping blocks and a mold surface. The radius of the particleboard panel determines the final size of the jamb.

I watched craftsmen Roger Paul and Jerry Mullis make the frame. They started with a given finished radius (the inside of the finished window frame) of 73 in. Based on previous trials, they estimated springback to be 2½ in., so the jig had to have a radius of 70½ in. By subtracting the thickness of the mold surface (¼ in.) and the depth of the clamping blocks (3 in.), they came up with a panel radius of 67¼ in. A new panel must be cut specifically for each size window frame, but smaller panels can be cut from larger ones to save on materials.

Paul and Mullis propped this panel vertically on a bench, and screwed to its rounded top a set of clamping blocks at about 8 in. o. c. Made from scraps of ¾-in. plywood laminated in pairs, these blocks are saved for reuse in each new jig. Finally, Paul and Mullis stapled a skin of ¼-in. hardboard, as wide as the intended window jamb, to the clamping blocks. They started at one end and carefully worked their way to the other, keeping the skin centered on the blocks and checking the final jig for bowing. Because any imperfections in the skin would telegraph through to the window jamb during glue up, Paul and Mullis checked that the skin was perfectly smooth. As a final touch, they waxed the skin with paraffin and smoothed the wax with steel wool. The wax keeps glue from sticking to the skin, so the finished jamb will pop easily out of the jig.

Assembling the plies—Curved jambs are built from four plies, each ³⁄₁₆ in. thick, about ¼ in. wider than finished width to allow for edge-jointing later, and a few inches longer than arc length. Paul and Mullis first sorted among the four plies and selected the best for the inside finished surface. They finish-sanded this face—a job that is much easier on a flat bench *before* glue up. They laid that ply face-down on the bench and on its back they drizzled aliphatic glue, each man releasing his artistic talents in a distinctive pattern of squiggles. They spread the glue to an even film, paying special attention to coating the surface along the edges. Then, they lifted the second ply onto the first and repeated the procedure until all four plies were glued. They clamped the plies by the edges to keep them together as a unit until clamped in the jig.

Clamping—Working quickly before the glue set, Paul and Mullis lifted the group of plies onto the jig and balanced it on the apex. Starting at the center, they laid a caul across the width of the plies, centered over a clamping block. They hooked one bar clamp under the chin of the block, snugged it down and then did the same with another clamp on the opposite side. Working as a team, one on each side of the jig, Paul and Mullis worked away from the center, clamping to each block with just enough pressure to hold everything in place. With the frame now held stable in the jig, Paul and Mullis went back and added two C-clamps, with a caul above and below, between each of the clamping blocks (drawing below). With all the clamps in place, they tightened them firmly, and glue flowed from every seam. They have found that only an even clamping pressure will give a fair arc. The key is to use lots of clamps and tighten them uniformly. The other men in the shop joked about Paul and Mullis hatching another peacock (photo below), and in fact, the radiating bar clamps did call to mind one of those strutting birds.

I returned the next morning to watch the plucking of the bird. Clamps were removed, scraped free of dried glue, and hung back in their racks. With the bench clear, Paul and Mullis laid the finished frame against the penciled layout marks. The frame had sprung to within ⅛ in. of the design width.

—J. Azevedo, a free-lance technical writer in Friday Harbor, Wash. Photo by the author.

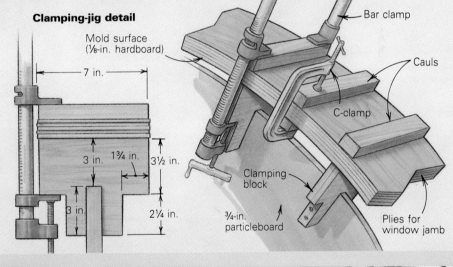

Clamping-jig detail

Mold surface (⅛-in. hardboard)

7 in.

3 in. 1¾ in. 3½ in.

3 in. 2¼ in.

¾-in. particleboard

Bar clamp

Cauls

C-clamp

Clamping block

Plies for window jamb

Raised Paneling Made Easy

A combination of built-up moldings and layers of plywood duplicates a traditional look

by Jim Donnelly

Glenn Bostock, a cabinetmaker living in Pipersville, Pennsylvania, was contracted to fabricate a frame-and-panel wall in the entryway of a house in New Hope, Pennsylvania. A formidable task, but certainly not an insurmountable one. There was, however, a challenge: The paneled wall was to be built on the inside curve of a circular staircase (photo facing page).

Traditionally, curved frame-and-panel walls were made by sawing, shaping and planing solid wood. Even with a fully equipped modern shop, the stiles, rails and panels would require elaborate jigs and extensive shaper work.

Bostock came up with a technique for duplicating the appearance of traditional frame-and-panel construction that was less expensive and quicker than traditional methods. His technique employs a series of overlapping layers, including bending plywood, tempered hardboard and kerfed pine. Each layer was nailed or glued in place, and the whole assembly then was trimmed out with moldings that hid the seams and fasteners of earlier layers.

Bostock's process was simple, and the most difficult task was making accurate cuts so the molding fit together well. Granted, the construction of the wall took a lot of time, but not nearly as much time as a solid-wood raised-panel wall would have required.

Although Bostock's project was a curved staircase wall, his techniques would work just as well for building a flat raised-panel wall. An added benefit for those with limited access to a shop is that Bostock's techniques can be accomplished almost entirely on site.

Two layers of bending plywood provide a firm foundation—Bostock began by attaching two layers of ⅜-in. bending plywood directly to the studs. Bending plywood is made of three laminations; the middle layer is thinner than the two outside layers. Eventually, the two layers of bending plywood would be covered by other layers of panels, stiles, rails and moldings. The bending plywood provides a solid base upon which to attach subsequent layers.

Bostock attached the first layer of bending plywood with nails and construction adhesive. The construction adhesive filled any gaps between the plywood and the studs, and the nails were more than adequate for drawing the plywood up to the studs. The second layer was put on with

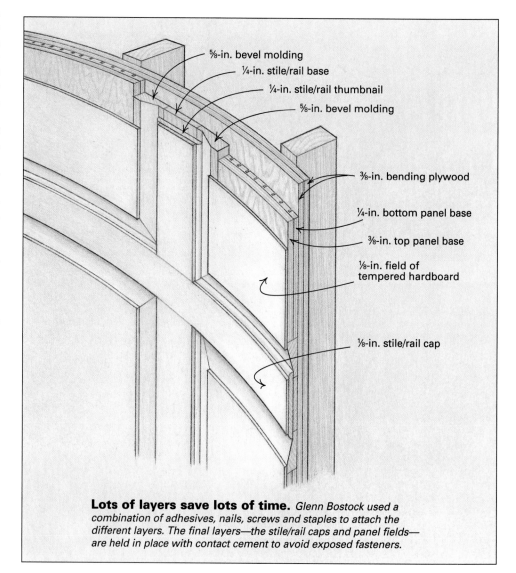

Lots of layers save lots of time. *Glenn Bostock used a combination of adhesives, nails, screws and staples to attach the different layers. The final layers—the stile/rail caps and panel fields—are held in place with contact cement to avoid exposed fasteners.*

⅝-in. bevel molding
¼-in. stile/rail base
¼-in. stile/rail thumbnail
⅝-in. bevel molding
⅜-in. bending plywood
¼-in. bottom panel base
⅜-in. top panel base
⅛-in. field of tempered hardboard
⅛-in. stile/rail cap

carpenter's glue and 1¼-in. staples. The yellow glue turned the two pieces of the bending plywood into one unit. Bostock used a paint roller to apply the glue quickly. The staples were placed every 6 in. to assure a tight, even fit between the two layers.

Lay out the frames and panels—Once the bending plywood was attached, Bostock marked the stiles, rails and panels on the wall. Bostock

colored in the location of the rails and stiles with pencil to make them easy to identify. Bostock drew all vertical lines with a level, and he then measured horizontal distances off a level line close to the floor (top left photo, p. 80). This horizontal line was established by marking points in several locations around the radius with a level and connecting the points using a thin piece of wood bent into the curve. Once the layout of the vertical and horizontal lines was complete,

Drawings: Vince Babak

The traditional look is made of nontraditional methods. By building up layers of different materials, including bending plywood, pine moldings and tempered hardboard, Glenn Bostock duplicated the look of a traditional frame-and-panel wall.

Bostock began the installation of the stile and rail base pieces.

The first layer of the stiles and the rails must be exact—In a traditional frame-and-panel wall, each stile and rail is made of one piece of solid wood. Each stile and rail in Bostock's wall is made of three pieces: the base; the thumbnail; and the cap (drawing facing page). The ¼-in. by 2½-in. clear, sugar pine base pieces provide the

foundation to which all the components are attached. All of Bostock's other layers of paneling either butt against or lie on top of the stile/rail base pieces.

Attaching the base pieces was simply a matter of nailing or screwing them to the bending plywood. But Bostock stressed that this point was the last chance he had to make sure everything was level, plumb and square, and because subsequent layers would be easier if the corners to

be mitered were 90° instead of 91° or 89½°, he took extra time to double-check everything (bottom left photo, p. 80).

Although the layout of the stile and rail base pieces was critical, the butt joints where the pieces meet was not nearly as critical because the thumbnail pieces and the caps would cover the butt joints. For this curved-wall project, ¼-in. thick pieces of wood flexed enough to bend easily around the 8-ft. radius of the wall's curve.

Pencil in the locations of the frames. Bostock marked out the stiles and rails on the wall. All subsequent layers of the wall either will butt against or be attached to the stile/rail pieces.

Accurate placement of the base pieces expedites the project. Making sure that the stiles are plumb and the rails level will ensure that the miters of the bevel moldings layers are true 45° cuts and that the panels pieces have 90° corners. Along the upward sweep of the staircase, some of the angles had to be bisected (sidebar facing page).

Miters have to be exact. Butt joints don't. Miter joints where the bevel moldings meet will be visible on the finished wall. Long butt joints between the bevel moldings and the stile/rail base pieces will be covered by the layers of the panel field. Screws used to attach the bevel molding also will be covered by the field.

Fasteners are hidden by gluing on a final layer of hardboard. One-eighth-inch tempered hardboard is used for the panel fields. Rolling on contact cement to the fields and sticking them to the wall covers all the staples, screws and nails Bostock used to attach the sublayers of the panels.

Bostock glued and stapled the stile and rail pieces into place.

Bevel pieces mimic the edge of a raised panel—There are two parts to a raised panel: the field and the bevel. The field is the large, flat section in the center of the panel, and the bevel is the diagonally cut border around the field. The panels in Bostock's project are made of four pieces: two panel bases; the field; and the bevel molding (drawing, p. 78).

Bostock made his bevel molding on a shaper using a panel-raising cutter. A similar profile could be cut on a router table. (For more information on using a router table, see *FHB* #90, p. 61.) A table saw also can be used to make the pieces if a shaper or a router is unavailable.

Even though the width of the bevel moldings is somewhat arbitrary, the thickness is crucial. The molding must be exactly as thick as the two layers of the panel base pieces, which in this case totaled ⅝ in. More on the panel bases later.

Bostock milled his bevel molding stock to a width about 2½ times as wide as needed. He ran the boards through the shaper twice, raising both edges. Then he ripped the moldings to their desired width. This method allowed him to run wider stock through the shaper more safely.

As with all the previous pieces in Bostock's project, the bevel moldings must bend. But the moldings are relatively thick and unyielding and had to be kerfed to ensure flexibility. Bostock laid the moldings so that the diagonal cut was face down on a power miter box. He then cut a series of kerfs into the thickest part of the back of the moldings.

Mitering the bevel moldings—The bevel moldings had to be mitered to fit within the framework created by the stile/rail base pieces. Because of the geometry involved in flexing the beveled molding into place, some gaps were inevitable between the bevel moldings and the framework made of the stile/rail pieces. These gaps would be covered by subsequent work (top right photo, facing page). The miter joints of the bevel moldings, however, had to be tight.

On Bostock's curved wall the run of the staircase creates a series of triangular panels, and fitting the bevel moldings into these frames posed greater challenges because their tight angles were not 45° miters. Rather, they were more acute and irregular. In order to fit the pieces accurately, Bostock had to bisect all the angles individually (sidebar right).

Because of the difficulty presented by their curvature, Bostock fit and installed the horizontal members of bevel molding first. He used yellow glue and 1¼-in. staples to fasten the moldings, but some pieces required screws to pull them up tight. The fasteners were kept close to the outside edge of the moldings so that subsequent layers would cover them. After installing all the horizontal pieces, Bostock installed the verticals.

Installing the panel base pieces—The panel base pieces are made of, respectively, kerfed ¼-in. plywood and ⅜-in. bending plywood. Both are used to fill the interior area created by the bevel moldings. Keep in mind that for a flat wall where the panels didn't have to bend, you would be able to use single pieces of ⅝-in. plywood to make the panel bases.

Bostock wasn't too concerned about getting a tight fit between the panel bases and the bevel moldings because the panel field pieces would cover the joint.

The stile/rail thumbnail molding must be coped—The stile/rail thumbnail moldings cover the joints between the stile/rail base pieces and the bevel moldings. Bostock made the stile/rail thumbnail moldings of ¼-in. pine. He milled a quarter round on one or both edges, depending on each piece's location in the project.

Bostock installed the perimeter pieces of the thumbnail molding first. Then the horizontal pieces, or rails, were installed. Rail ends had to be coped into the perimeter stiles.

On Bostock's curved wall, to fit the curved rails, the angle-and-cope cut was made on one end while the other end ran long over the perimeter stile. He temporarily screwed the rail into place, which allowed for a direct marking of the rail in relation to the stile. Once marked, he removed the screws and routed the cope with a ¼-in. cove bit in a router table. Then he reinstalled the rails using glue and the same screw holes.

After all the rails were in place, Bostock coped and fit the rest of the stiles. Because these pieces are coped into the rails, their length is determined by measuring the distance between rails and adding ¼ in. for each cope. Extra wood is removed during the course of coping.

Stile/rail caps were attached with contact cement—Bostock made the last piece to cover the rails and stiles—the stile/rail cap—of ⅛-in. sugar pine that was planed in his shop. If you don't have a planer, you could use ⅛-in. hardboard, which is available at some lumberyards.

Mitering odd angles

Mitering wood at an odd angle is exacting work, but it's not hard if you have a bevel gauge and a compass. The process involves bisecting the angle to be mitered.

The first step is to use your bevel to transfer the angle to be mitered onto a scrap of wood or a piece of paper. Next, stick your compass point in the apex of the angle and strike an arc equidistant from the apex along both legs of the angle (drawing below). The distance at which you strike the arc is arbitrary, but it is probably a good idea if you set your compass so that the marks you make are at least 3 in. from the apex.

The next step is to position your compass at the point where the first arc intersected the angle's leg. Draw a new arc through the approximate center of the angle. Repeat the process on the other leg. Again, the distance between the metal point and the pencil point on the compass is arbitrary, but the setting has to be at least wide enough so that the two arcs will intersect.

The intersection point of the two new arcs establishes the center of the angle. Now you can reset your bevel to measure the bisected angle from the apex to the point where the arcs intersect. Transfer the new angle to the pieces of wood that are to be mitered.

Keep in mind that there is plenty of room for error in this process and that it might behoove you to cut your pieces a little long so that you can fine-tune the miter with a block plane.—*J. D.*

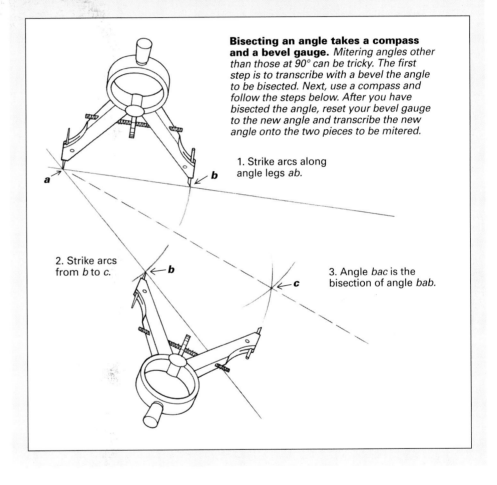

Bisecting an angle takes a compass and a bevel gauge. *Mitering angles other than those at 90° can be tricky. The first step is to transcribe with a bevel the angle to be bisected. Next, use a compass and follow the steps below. After you have bisected the angle, reset your bevel gauge to the new angle and transcribe the new angle onto the two pieces to be mitered.*

1. Strike arcs along angle legs *ab*.

2. Strike arcs from *b* to *c*.

3. Angle *bac* is the bisection of angle *bab*.

Because nail holes or other mechanical fasteners would detract from finished work and because there was no easy way to clamp the pieces for a conventional glue joint, Bostock used contact cement to install the stile/rail caps.

The panel fields are made of ⅛-in. hardboard—The panel fields are the final pieces that Bostock installed on his paneled wall. These pieces cover both the fasteners he used to attach the outside panel-base layer and the joint between the panel base and the bevel moldings.

Bostock made his panel fields of ⅛-in. tempered hardboard (bottom right photo, facing page). The material bends well around the curve

of the wall, and its smooth surface takes paint readily. Like the stile/rail covers, Bostock attached the panel fields with contact cement. The cement can be brushed on or rolled on; Bostock prefers the latter method. Rolling provides for an even coating of cement, critical to a good bond.

When Bostock was finished with the woodwork of the wall, he applied latex caulk and glazing compound to fill any gaps between layers that would otherwise reveal that this project was not conventional raised-panel construction. ☐

Jim Donnelly is a cabinetmaker and writer who lives in Bucks County, Pennsylvania. Photos by Glenn Bostock.

Raised-Panel Wainscot
Traditional results with table saw and router

by T. F. Smolen

Installing a traditional raised-panel wainscot is a good way to transform a nondescript room into a more formal space. It's also a handsome alternative to replastering old walls that have been damaged over the years by feet, furniture and children. A wood wainscot is more durable than plaster or gypboard, and it relieves the unbroken plane of the wall with the delicate array of shadow lines created by moldings and flat surfaces, as shown in the photo at the top of the facing page.

The term "wainscot" is often loosely applied to various paneling treatments that cover the lower part of a wall. The raised-panel wainscot shown here is a traditional style based on frame-and-panel construction. The frames consist of vertical members, called stiles, and horizontal members, called rails. They support panels whose beveled borders and raised fields give this particular style its name. Each panel rests in its frame with its grain running vertically. Panel width is limited by the width of your stock, unless you edge-join two or more boards together.

Beneath the bottom rail of the frame, a baseboard extends to the floor. At the top of the wainscot, a molding called a chair rail covers the joint between the top rail and the upper section of the wall.

The moldings that I used in making this wainscot can be bought at most lumberyards, and it's possible to make the raised panels and their frames with a table saw. I used a slot cutter and router table to groove the inner edges of my stiles and rails, but you could handle these as well on a table saw with a ¼-in. dado blade.

Panel design—The wainscot that I installed in the dining room of my late Victorian home is traditional in design. I wanted its top to be about 41 in. from the floor. This finished height would include an existing 7-in. baseboard that could be left in place, and a 4-in. wide chair-rail molding that would overlap the top rail of the raised-panel frame. With the stiles and rails 4 in. wide, the panels would show 22½ in. of height. Their actual height would be 23½ in., since ½ in. of the panel edges would be let into the grooved frame all around (drawing, above right).

The width of the raised panels was determined by the distances between corners and the door, window and cabinet frames in the room. In each wainscoted section, I wanted

the panel size to be uniform, so I divided each section to allow equal spacings between stiles. The largest section has four panels, each of which is 12¾ in. wide; the smallest section has a single 18-in. wide panel.

I milled the panel stock from 4/4 roughsawn pine boards about 14 in. wide. The wood, originally intended for flooring, was air dried. I had one side planed down to a thickness of ⅞ in., which allowed for a full ⅛ in. of relief on the raised panel and a sturdy ¼-in. thick tongue around the panel perimeter. Tight knots were acceptable, because I planned to paint the finished wainscot.

Raising panels—I set up a cutting schedule that included all panels for the wainscot, crosscutting the planks to 23½-in. lengths, then ripping them to finished width (the distance between inside edges of the stiles plus a ½-in. tongue allowance on each side). Then I made a template of the panel profile, which I set against the sawblade or dado head when setting up for a cut in order to produce the proper bevel and depth.

Making a raised panel with a profile like the one shown in the drawing below left requires three cuts on each side—one to form the bevel, one to form the tongue on the panel's edge and one to form the shoulder on the edge of the field.

To cut the bevel, I used a carbide-tipped combination blade because it produces a smooth surface that needs little sanding. To set up the saw for the bevel cut, I first set the arbor angle and fence distance to match the bevel on the template. Then I clamped a guide board to the tabletop, parallel to the fence and ⅞ in. away from it. This guide board aligned and steadied the on-edge workpiece as it was fed through the saw (photo facing page, center). Without it, you'd have a troublesome, hazardous time keeping the bevel straight and true. Even with the fence, the blade had to cut through just over 2½ in. of wood, so I held the stock securely and fed with slow, steady pressure. The next time I need a similar setup, I'll use a 4x4 as the auxiliary fence.

Because some boards cup slightly after they are surfaced, I found it best to make cross-grain bevel cuts soon after the boards had been cut to their finished sizes.

When the bevels had been cut on all four sides of the panel, I completed the panel pro-

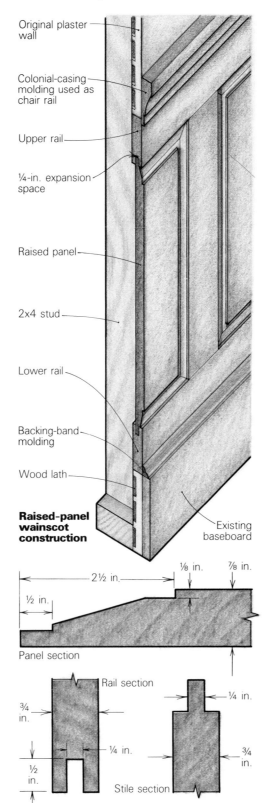

Original plaster wall

Colonial-casing molding used as chair rail

Upper rail

¼-in. expansion space

Raised panel

2x4 stud

Lower rail

Backing-band molding

Wood lath

Raised-panel wainscot construction

Existing baseboard

2½ in. · ½ in. · ⅛ in. · ⅞ in.

Panel section

Rail section · Stile section · ¾ in. · ¼ in. · ¼ in. · ¾ in. · ½ in.

Stile is tenoned into upper and lower rails.

file using 1-in. wide planer knives (flat across the top) mounted in a Sears molding head.

The second cut (photo bottom left) removes a triangular section of waste at the edge of the panel to create the ¼-in. thick tongue that fits into grooves in the frame. To allow for expansion and contraction of both frames and panels, I trimmed ⅛ in. from the top tongue of each panel and ³⁄₁₆ in. from each side tongue (wood expands more across the grain than along it). The inner edge of the frame would be grooved to a depth of ½ in. to provide expansion space at the top and sides of each panel. Allowing for play in the fit of each panel in its frame is necessary if the wainscot is to survive years of fluctuating humidity. Too tight a fit, and the panels are likely to check or bow out of their frames.

The third cut produces a ⅝-in. wide land at the juncture of bevel and field, and a ⅛-in. high shoulder where the field begins, 2½ in. from the panel edge. No auxiliary fence is required for this cut, but the main fence needs to be set up exactly right. Here again I use the template to get accurate settings for the fence and blade. The blade should just graze the wood surface.

As soon as the panels were cut, I prefin-

In a traditional raised-panel wainscot, a solid wood panel sits in a grooved frame. A chair-rail molding covers the joint between the top rail and the plaster. Overlapping the bottom rail, a baseboard extends to the floor, as shown in the drawing on the facing page.

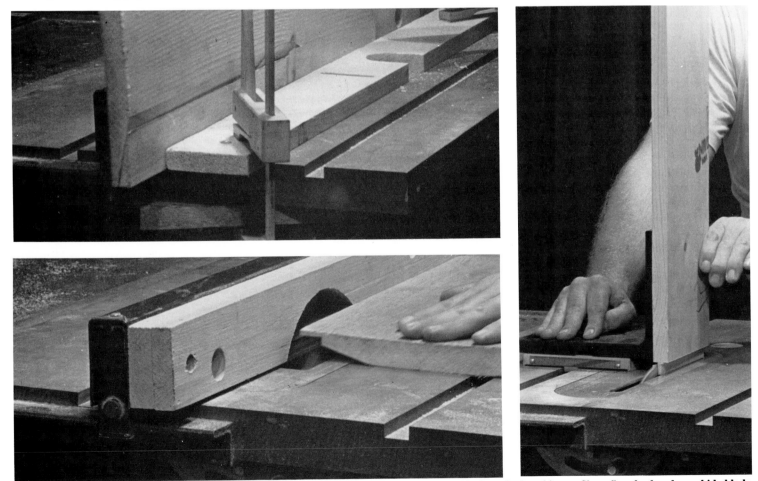

Table-saw setup. **Top, a guide board, clamped parallel to the main fence, aligns the panel blank as the bevel is cut. Slow, firm feed and a carbide blade produce a smooth cut. For safety, let the blade stop spinning before the waste piece is removed. Above, a 1-in. wide planer blade cuts the tongue along the edge of the panel. At right, the author uses a tenon-cutting jig, attached to the saw's miter guide, to cut the tenons on a stile.**

ished them to keep checking, cupping and wood movement to a minimum while I built the frames. I filled dings and small knots with plastic wood; then I sanded the exposed face with 80-grit paper and sealed knots with shellac to keep sap from bleeding through the finish coats of paint. Finally I gave each panel a coat of oil-base sealer compatible with the enamel finish I planned to use.

Stiles and rails—First I ripped 1x6 pine boards to 4-in. width. I cut all rails about 3 in. longer than their finished length so that after assembly I could scribe them for an exact fit to the walls on either side.

My design called for a ¼-in. wide by ½-in. deep mortise-and-tenon joint between stile and rail, so I grooved the inner edges of all the stiles and rails on my router table, using a ¼-in. slot cutter. Since the slot has to be cut down the exact center of the stock, I tested the setup on ¾-in. thick scrap before running stiles and rails through the machine.

To cut the tenons on the stiles, I used a thin-rim carbide blade on my table saw and a tenon-cutting jig on my miter gauge (photo previous page, bottom right). After a little touchup work with the chisel, I was ready to put frames and panels together.

Assembly and installation—I clamped the bottom rail of each panel section in the end vise of my workbench and fit each section together dry. Small pencil lines on panels and rails served as registration marks for centering each panel in its frame.

In assembling the wainscot, only the stile tenons get coated with glue. The panels are seated firmly in the frame, but not glued. This way, they can respond to changes in humidity and temperature without binding or bowing their frames. I used a small brush to spread glue on the tenons, and then assembled the frames and panels, snugging stile-to-rail joints together with pipe clamps until the glue set.

I wanted to nail the wainscot directly to the studs rather than installing it over the existing plaster, which was sound but presented quite an irregular surface. Leaving the original baseboard intact, I snapped a level line about ¼ in. above the installed level of the top rail and ripped out the plaster and lath. I scribed the plaster along this line with a utility knife, then ripped it off by hand. I cut the lath with a chisel and pried it off the studs with a hammer and small prybar.

Next, I nailed up the paneled sections, which I had purposely built slightly wider than the spaces they would occupy so they could be scribe-fitted to the walls. This left a slight gap between baseboard and bottom rail and between plaster and top rail. I used a 4-in. wide Colonial-casing molding at chair-rail height to cover the joint between the rough plaster and the top rail of the wainscot, and a 1⅝-in. wide backing-band molding at the baseboard and bottom-rail junction. □

Ted Smolen practices law and does amateur woodworking in Danvers, Mass.

Modified wainscot: a raised panel with birch-veneer plywood and beveled molding

by Michael Volechenisky

A small ad in the local paper got me interested in building a raised-panel wainscot. It offered for sale "paneled wainscoting from a 150-year-old home" (wainscot is actually the correct term here). This would be just the thing for the dining room in my equally old house, which I was in the middle of redoing. But the age of the wainscot was unfortunately confirmed by its condition, and in any case there wasn't enough of it to go around my 15-ft. by 15-ft. dining room. So I decided to build my own from scratch, enlisting the advice of Luther Martin, a retired builder and woodworker who was sympathetic to the idea of recreating an old look with new materials.

Panel design—I planned to construct a frame-and-panel wainscot along traditional lines, bordering it with a baseboard along the floor line and a chair-rail molding along the top. But Martin didn't want to use solid wood panels because he'd seen long stretches of wood-paneled wainscot push walls out of plumb as a result of normal wood expansion. Cracking and cupping are other risks of solid wood panels. So I made my raised panels from ¾-in. thick lumber-core plywood with birch face veneers, and with slightly modified beveled molding. The molding, which is 2 in. wide, forms the beveled border of the panel, and the lumber-core plywood is the field. A tight tongue-and-groove glue joint between border and field and two coats of white enamel hide the fact that these panels weren't raised in the traditional fashion.

This alternative design has several advantages. First of all, the lumber-core plywood field is far more stable than its solid wood counterpart. Expansion and shrinkage are negligible, as are cracking and checking. And the birch face veneer is better than solid pine or fir if you're planning to paint—as I was—because it contains no knots or resin pockets, which could bleed through the finish.

Third, you can give the panel's bevel a fancier treatment than is possible with conventional techniques, since the border isn't an integral part of the field. The molding I used, for example, has a quirk bead at the inner edge of the bevel—an embellishment that suggests far more intricate work than was actually involved.

Michael Volechenisky lives in Sayre, Pa., and Pompano Beach, Fla. Photo by the author.

Making moldings—From top to bottom, my wainscot contains a chair-rail molding scribed to fit the plastered wall; two smaller moldings (a cove and a beaded stop) that fit over the frame and raised panels; and a baseboard scribe-fitted to the floor, with its top edge covered by a modified scotia molding (drawing, facing page). I made these moldings with my spindle shaper, but could have bought similar ones at the lumber store.

I milled the beveled molding for the panels that make up the wainscot on an old Hebert molder-planer. The Hebert, which is no longer made, is similar to the Williams & Hussey planer (Williams & Hussey Machine Co., Dept. 16, Milford, N. H. 03055), and both machines are shop-size versions of the larger, more powerful planer-molder machines used by lumber mills.

My machine has only one cutterhead, which is mounted horizontally above the table and holds a pair of knives. It is driven by a 1-hp motor, and is powerful enough to complete a molding in a single pass, providing you use knot-free wood that's not too dense. But to be on the safe side, I usually make my first cut to within 1/16 in. of the finished dimension and then run the stock through a second time to get a smooth surface that requires very little sanding.

Molder-planer manufacturers sell a variety of molding cutters to fit their machines, but I've often made my own from precision-ground tool steel (it comes in many sizes and thicknesses, and can be bought wherever metalworking tools are sold). After drilling two holes in each tool-steel blank so they can be bolted to the cutterhead, I transfer my molding outline to the blank and start removing metal. I hacksaw as much as I can, grind the shape to a 30° bevel and hand-file corners and coves where my bench grinder can't reach. After getting both cutters as nearly identical as I can, I mount them in the Hebert and run a trial piece of wood through. This tells me which cutter is doing most of the work by the flecks of wood that adhere to its cutting edges. More filing follows; then I hone the blades and install them. Since most of my molding runs are for 500 ft. or less, it doesn't seem necessary to harden my cutters.

All the moldings, stiles and rails for my wainscot were cut from basswood stock I'd been saving. Basswood works easily, and I've found that you can usually smooth it by hand with a cabinet scraper, with little or no sanding. It also takes a fine coat of paint, because it's knot-free and resin-free.

No matter what type of wood you use on a molder, you'll get better results if you make sure that each board you run through the machine has its grain oriented correctly relative to the cutterhead. To prevent small chips and tears in the molding, feed your boards so that their grain slants down toward the exit side of the machine.

I constructed the panels first. Each completed panel actually consists of five pieces—the birch-veneered lumber-core field and four bordering molding sections. As shown in the drawing, the tongue along the inner edge of the molding is designed to fit in a ¼-in. wide groove cut in the edges of each panel. I grooved the panel edges on my table saw, using dado blades. At the corners of the panel, adjacent molding sections are mitered. Once all the parts for a panel were cut and test-fitted, I glued them up.

After the panels were finished, I built their frames. Top and bottom rails for each 15-ft. side of the room were cut from 16-ft. long basswood boards. Using a hollow-chisel mortiser chucked in my drill press, I cut mortises in the rails to receive stile tenons. Then I grooved the inner edges of the frame on the table saw to receive the ¼-in. by ¼-in. tongue around each panel.

I test-fit the frame-and-panel sections for each side of the room, then glued and clamped them together. Before installing each section, I gave the back of the frames and panels two finish coats—the same number that the front of the wainscot would receive. Thus both sides of the wood can respond equally to temperature and humidity. Perhaps this wasn't necessary, but there hasn't been a single paint crack in the wainscot after eight years on the wall.

Though I could have installed the wainscot directly over the dining room's old plaster walls, they were in such bad shape that I stripped the room down to its studs and nailed up rock lath. Then new plaster was applied down to a temporary ground I nailed just below the height of the chair-rail molding that would top off the wainscot.

The frame-and-panel assembly was the first part of the wainscot to get nailed up. I made sure each wall section was level and used 8d finishing nails, positioning them close to the rail edges so they would be hidden by the covering layer of molding.

Once the four frame-and-panel sections were up, I added moldings to the rails. The baseboard and the scotia-style molding covering its top edge were mitered at the corners, as were the chair rail and its two adjacent moldings. A light sanding, followed by a primer coat and two coats of semi-gloss Kemglow enamel, finished the job. □

The gluing setup used to construct the panel consists of four bar clamps and a Formica-faced base slightly shorter and wider than the panel. At panel corners, the molding joint is a glued miter.

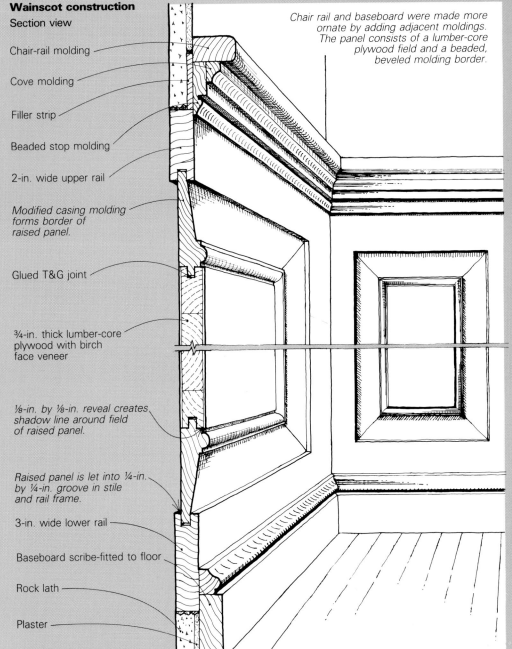

Wainscot construction
Section view

Chair-rail molding

Cove molding

Filler strip

Beaded stop molding

2-in. wide upper rail

Modified casing molding forms border of raised panel.

Glued T&G joint

¾-in. thick lumber-core plywood with birch face veneer

⅛-in. by ⅛-in. reveal creates shadow line around field of raised panel.

Raised panel is let into ¼-in. by ¼-in. groove in stile and rail frame.

3-in. wide lower rail

Baseboard scribe-fitted to floor

Rock lath

Plaster

Chair rail and baseboard were made more ornate by adding adjacent moldings. The panel consists of a lumber-core plywood field and a beaded, beveled molding border.

Drywall Detailing

An alternative to wood trim around doors, windows and skylights

by Dennis Darrah

Trimming out doors, windows and skylights can be a rewarding endeavor. It can also be costly, if done with the proper care and materials that this labor-intensive procedure requires. Unfortunately, as a house nears completion and costs are running over estimate, this can be one area of a job that experiences severe cutbacks. The most frequently used alternative to first-rate trim is inexpensive pine, applied as plainly as possible. Many owner-builders, and even some contractors, justify this route with the argument that one can go back through the house when time and money allow, and redo the trim in proper fashion. Of course, we all know that the chances of this happening are slim.

A more graceful alternative to the pine solution is drywall detailing around windows, sky-lights, archways, and even doors. Although the labor involved is comparable to that for installing wood trim, the materials are much less expensive. Don't get me wrong—I love beautiful trim work. But I also appreciate a project brought in on budget. And even where money isn't the overriding factor, drywall detailing may still be the most effective and pleasing solution to many trim problems. This is particularly true in renovation, where old work must abut new.

Mud and metal—The basic materials for detailing wall openings are the various beads, moldings and tapes that are available through a good drywall supplier (drawing, p. 88). If there isn't a good drywall supplier in your area, one good mail-order source is Bon Tool Co. (4430 Gibsonia Rd., Rte. 910, Gibsonia, Pa. 15044; 412-443-7080). I have not used the blueboard and veneer-coat system of plaster, so here I'll concentrate on the standard drywall accessories.

Corner bead is the obvious choice for forming wall returns into windows, without the use of casings. It's available in galvanized steel and white vinyl. Corner bead can be attached by driving a drywall nail every 4 in. to 6 in. along each side of its length. The length of the nail will depend on the thickness of the drywall; the nail should penetrate the framing ⅜ in. to ½ in. One advantage of metal corner bead is that you can attach it with a crimping tool, which crimps the edges of the bead so that it grips the drywall. This saves quite a bit of time, especially on large jobs, and also helps

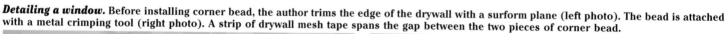

Detailing a window. **Before installing corner bead, the author trims the edge of the drywall with a surform plane (left photo). The bead is attached with a metal crimping tool (right photo). A strip of drywall mesh tape spans the gap between the two pieces of corner bead.**

to center the bead. If the bead is rolled to one side, it makes the other side difficult to cover with compound. Though I use a crimping tool to set the bead, I always add four to six nails for extra security. Any structural movement, particularly in a new house, is liable to show around windows and doors, and the first casualty will always be the drywall work. But there are times when nails can't be used, particularly in old houses where the interior walls have been covered with a layer of foamboard insulation beneath new drywall. The foam and drywall create built-up corners through which nails won't easily reach. In such cases, a crimping tool is all one can use in applying standard metal corner bead. Crimping tools retail for $60 to $70.

There is an alternative to the rigid metal or vinyl corner bead: a paper tape lined with two strips of thin galvanized steel, which forms a rigid corner when folded lengthwise (photo next page). It's sold under the names Flex-Corner and Sure-Corner, and it comes in widths of 2¼ in. to 4 in. This product is especially useful on oblique or acute angles, the kind that you're likely to find in skylight openings. It is designed, like standard paper tape, to be embedded in a layer of mud and then mudded over. It is folded at the center margin and applied with the metal facing the wall. In my estimation, it is not quite as serviceable as rigid corner bead; it doesn't always adhere well, it occasionally has to be reworked and it doesn't

Photo: Andrew Kline

A stool cap and some paint are the only trim-work materials needed around this window.

stand up as well to daily abuse. Despite this product's limitations, there are times when it is the only option.

Where the drywall meets the window sash, you have two alternatives: J-bead or L-bead. The basic difference is that J-bead shows as the final detail, while L-bead is mudded over (drawing next page). J-bead must also be installed when the drywall is installed. Which one I specify depends on the type of track in which the sash is mounted. J-bead works well for movable sash—window movement might cause the mud needed for L-bead to crack over time. L-bead works better for windows with complicated track where J-bead can look too busy. However, J-bead actually enhances simple retrofit track, creating a handsome thickening effect at the juncture of the return and the sash. Although you should feel free to use one or the other based on your own aesthetic, there is one caveat: if you expect high condensation levels on the window or skylight in question, stick with J-bead. Your mud work over the L-bead is liable to be damaged by the moisture collecting on and around the window.

I try to have the painters spray the J-bead before it's applied. Spray paint adheres better to metal than does a brushed-on coat. An ideal alternative, at least for white walls, is to use white vinyl J-bead, but this is usually hard to find in small quantities. I also recommend that moisture-resistant drywall be used for all

Three coats of joint compound are applied with a 6-in. knife (left photo). The opening is sanded with 120-grit, open-weave silicon paper (right photo). No sanding is done until after the final coat has been applied.

returns around windows and skylights that have potential condensation or moisture problems.

Windows are not usually that tricky or time-consuming to finish (bottom photos, previous page). On most returns one can usually catch both the L-bead and the corner bead in one pass of a 6-in. knife. If J-bead is used, all you have to finish are the outside corners, though on large houses this can add up to quite a bit—keep that in mind when pricing out the job.

I finish skylights much like I finish windows. First I apply the corner bead to the square corners and the flex tape to the oblique angles. I use the mesh tape on the inside corners—in my experience it holds the compound better than paper tape. I also use mesh tape to span the joints between two pieces of corner bead or a piece of corner bead and a piece of flex tape. This is necessary because the corner bead and flex tape can't overlap. Their combined thicknesses would result in a lump that no amount of tape would cover. For the first coat, I tape the inside corners, let them dry, then tape the outside corners. That way I'm not constantly fighting overlapping knife marks. I use three coats of compound. Less compound is used in the second and third coats, and because overlapping marks are less of a problem, those can be done all at once. I don't usually sand until after the final coat is done. I sand with 120-grit, open-weave silicon paper; it cuts better than standard paper, and it doesn't load up with compound as fast, either.

Be aware, however, that skylights take time to do right. If that isn't bad enough, your taping work is under the scrutiny of the most severe lighting possible—direct sunlight. To look good, it has to be done well. A bad taping job is the worst of all possible finishes.

Levels of perfection—I do both custom homes and commercial work. If I were to choose only the biggest buck for the least

headaches, I would go strictly commercial; there are usually no ceilings to contend with, and most commercial spaces are lit with diffuse or fluorescent lighting, which is very forgiving of taping irregularities. I'm not downplaying the quality of commercial work, it just offers a different set of challenges than a custom home. On a commercial job, for instance, a good taper really has the chance to show off his speed.

In a home, on the other hand, your taping has to look good under wildly varying lighting conditions. And be forewarned: no amount of paint can hide a lousy taping job. Success depends not so much on how fast you can move, but rather on how fast you can work well—it's not speed alone, but speed combined with quality that counts. When I finish-sand a custom job, I hold a spotlight in one hand and a sander in the other. Around skylights, I try to sand on cloudy days. The light is more diffuse then, and the spotlight helps me to pick out minor irregularities that would be masked by strong sunlight. Or I sand at

An alternative to rigid corner bead: paper tape laminated to two strips of thin galvanized steel. It forms a rigid corner when folded lengthwise. Although not as sturdy as rigid bead, it's the only choice when taping oblique angles.

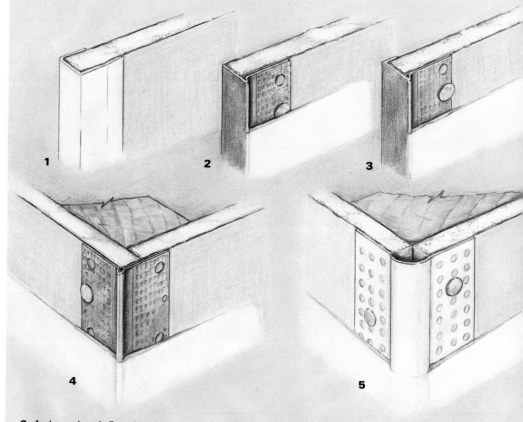

A bead sampler. There are a wide variety of drywall beads on the market. Although you'll find slight variations between manufacturers, the basic products are essentially the same. Below are a few of the more useful ones for trimming out doors and windows.

1. J-bead. Gives a finished edge without joint compound. Use it when the drywall edge must be isolated from a window sash or a door jamb—where there's a potential for condensation, or where door or window movement could crack the finished compound. Install it before applying the drywall. Galvanized steel J-bead can be painted before installation; white vinyl J-bead usually doesn't have to be.

2. L-bead. Leaves a crisp edge where the drywall meets the sash or the jamb. The exposed leg is finished with joint compound. Available in metal or vinyl.

3. U-bead. Gives a clean edge along with isolation of the drywall. Face nail and finish the same as L-bead. Usually installed before the drywall. Available in metal or vinyl.

4. Standard corner bead. Comes with a deep knurling for reinforcement of outside corners. Available in galvanized steel or white vinyl.

5. Rounded corners. Compound is applied to the legs; the nose is left bare and painted. Available in white vinyl or paper-faced galvanized metal.

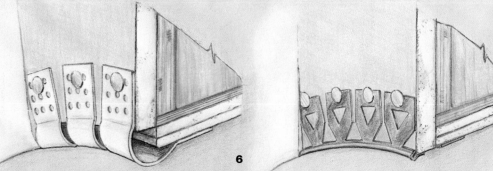

6. Archway bead. For trimming arched or rounded window and door openings. The white vinyl variety has a rounded edge; metal bead has a squared edge.

night, and check it out by the light of the next day. I find it hard to sand in the glare of direct sunlight and have it come out right.

Drywall can be dry-sanded, wet-sanded or sponged. There are arguments in defense of each method. I've never wet-sanded. I used to sponge, but I never got good results from it. The sponge always left visible ridges and smears that didn't seem to happen with dry-sanding. I now dry-sand and swear by it. The tools involved are hand sanders, pole-mounted or hand-held. The sanders are rubber-backed and take precut sandpaper (100, 120, or 150 grit) or Fabricut sandpaper, which is an open-weave silicon carbide sanding cloth that does not readily clog with dust. This comes in 100 and 120 grit. Power tools have no place in sanding seams. For one thing, you risk ripping the drywall paper with a power sander. Besides, if the seams are that heavily overloaded with mud, go back and do a skim coat or get out of the taping business.

I have only had one opportunity to trim out interior doors with drywall. The rough opening is wrapped with drywall, the corner bead applied, then the door jamb and the mud. A piece of cove molding helps ensure that no gaps will develop over time between the jamb and the drywall. Although the job really could be done without the cove molding, I feel the door opening looks a lot better when it's there.

Flat-taping—Being a timber framer in addition to a drywall taper, I've done a lot of flat-taping next to exposed timbers (photo below left); flat-taping is only necessary where stress-skin panels aren't used. I feel that it's impor-

tant to mask out the timber before applying the drywall. This not only speeds up the taping and painting, but protects the timbers throughout the whole process. Mud is applied to the wall abutting the timber, then paper tape embedded flat in the mud, its edge carefully aligned with the timber. The paper tape will cover any gaps between the timber and the drywall. If the gaps are excessive, they can be pre-filled with a patching plaster or even a spray foam.

A trick for speeding up the finish of a tim-

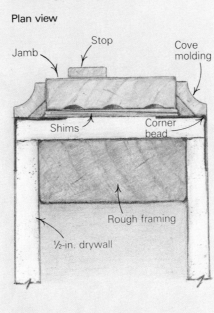

Detailing a door casing

Plan view

Jamb

Stop

Cove molding

Shims

Corner bead

Rough framing

½-in. drywall

ber-frame ceiling is to install the drywall from above. Simply prepaint the ceiling drywall and lay it on top of the timber joists, gluing and screwing it as you go. Lay the subfloor and finish floor on top of the drywall, then screw the drywall up into the subfloor. With typical 2-ft. centered timber joists, there is no ceiling taping to do whatsoever. To facilitate accurate screwing into the joists, pre-snap chalklines on the back of each piece of drywall. For beams with rough surfaces, you can protect the drywall by installing a strip of closed-cell sill sealer between the joists and the drywall.

I did one renovation job on an old timber frame with exposed rough beams on 4-ft. centers. Using the above system, I saved myself some 280 ft. of flat taping next to the edges of the beams. A word of caution, however—be careful not to step through the drywall before the subfloor is in place.

The bottom line—You're limited only by your imagination in the use of drywall to enhance interior finishes. For example, I once built drywall-formed light fixtures in a cathedral ceiling (photo below right), along with a false column to provide a wire-chase and to hide a structural steel rod.

Once you get good at doing drywall details you'll be in greater demand. You may find contractors and architects consulting you before completing the design of a house. If you work it right, you can get paid for your design suggestions, as well as for their execution. □

Dennis Darrah specializes in drywall finishing and timber-frame construction. He lives in Montpelier, Vermont.

A timber-frame alternative. Install the floor joists, then lay the drywall from above and avoid taping against the beams.

Photo: Dennis Darrah

Drywall-formed light fixtures are a graceful alternative to a truncated cathedral ceiling. Besides permitting overhead lights, they add interesting detail to the apex. Note also the drywall-formed wire chase beside the chimney.

Photo: Andrew Kline

Drawings: Bob Goodfellow

Building a Fireplace Mantel

Fluted pilasters and decorative trim make for a mantel that appears complicated but really isn't

by Gary Katz

At first glance, fireplace mantels seem impressively intricate and outrageously expensive, but frequently they aren't. I know because a good client asked me to build a copy of a mantel from his previous home. After seeing the original mantel and identifying its parts, I was able to build the new mantel simply and economically using solid stock, plywood and manufactured moldings (photo right). Even though I've been a trim carpenter for many years, anyone with a basic knowledge of woodworking can build a fireplace mantel with the techniques I used here.

Strips of MDF simplify flute spacing—For this mantel, the first things to build are the pieces flanking the firebox, or the pilasters. I make each pilaster with three pieces—two sides and one face—mitered together along their length. I rip the faces and sides from solid oak and cut them slightly longer than their finished length. Once the pilasters are built, I'll trim them to size on a radial-arm saw.

Although I rip the sides of the pilasters to width with one edge square and the other beveled 45°, I rip the faces slightly wide and with square edges. Later, I'll bevel the edges at 45° for mitered joints, but right now I'd rather push a square edge along the table saw's rip fence. The long point of a bevel can jam under the fence, and in this case I'll be pushing the face stock

Mantel built in three sections. A 2x4 prop (top photo) holds the mantel top level while the author nails a pilaster to the panel wall. The three components—two pilasters and a mantel top—were built off-site; the texture on the oak comes from scraping out the soft grain.

through the fence six times to make parallel decorative grooves, or flutes.

Although fluting looks ornate, it isn't difficult to accomplish with a table-mounted router. My Makita portable 8-in. table saw is designed with a router mount underneath, so the bit sticks up through an opening in the table. The fluting also can be cut by running a router along a straight-edge or by using a router guide. While these

methods might be safer, they are slower and less accurate than a table-mounted router.

I use a ½-in. round-nose (core box) bit to cut the flutes, set to a depth of ¼ in. To make the process easy, I place multiple strips of medium-density fiberboard (MDF) between the rip fence and the workpiece (top photo, facing page) and remove one strip after each pass, making it unnecessary to adjust the rip fence for each flute.

For the fluting on these 6-in. wide pilasters, I make the MDF strips 1-in. wide: ½ in. for the flute plus ½ in. for the space between flutes, which gives me even spacing across the face. The MDF strips stay put because they butt into a stop block that's screwed to the table extension. The stop block is positioned so that the workpiece bumps against it, resulting in flutes that end 1¼ in. from the bottom of each face.

On the MDF strips, I draw a line 1¼ in. beyond the feed side of the router bit; this line indicates the beginning of the fluting. I carefully lower the workpiece past the line for the initial plunge and slowly pull it back until the end of the stock lines up with the 1¼-in. mark across the MDF strips. Then I feed the board through until it butts into the stop block, remove the workpiece and an MDF strip and repeat the process five more times.

After fluting the faces, I rip the edges at 45°. Then I spread glue on the miters of the side pieces and the face, hold the joints tight with my

Side view shows combination of plywood and solid wood

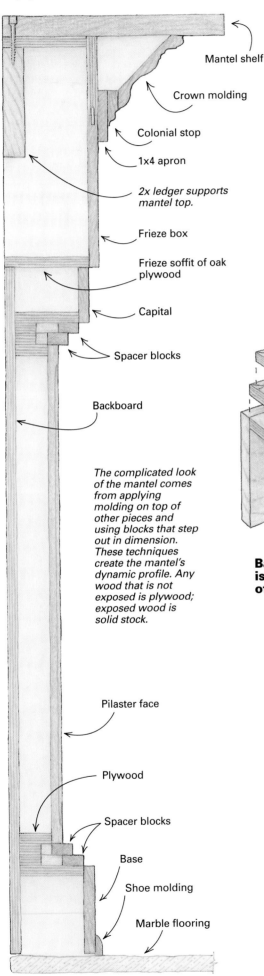

Mantel shelf

Crown molding

Colonial stop

1x4 apron

2x ledger supports mantel top.

Frieze box

Frieze soffit of oak plywood

Capital

Spacer blocks

Backboard

The complicated look of the mantel comes from applying molding on top of other pieces and using blocks that step out in dimension. These techniques create the mantel's dynamic profile. Any wood that is not exposed is plywood; exposed wood is solid stock.

Pilaster face

Plywood

Spacer blocks

Base

Shoe molding

Marble flooring

Cut multiple flutes without moving the fence. The pilaster faces were fluted on a router table by registering them against 1-in. MDF strips instead of directly against the rip fence. After each pass, a strip is removed; the rip fence is not moved.

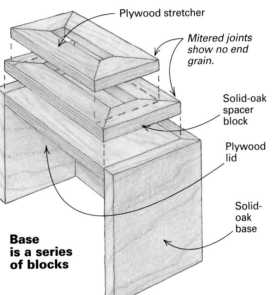

Plywood stretcher

Mitered joints show no end grain.

Solid-oak spacer block

Plywood lid

Solid-oak base

Base is a series of blocks

Joining the pilaster to the capital. After building up the capital with two 1-in. oak spacer blocks, the pilaster's plywood end is screwed to the capital. All of the blocks are flush.

Use plywood where it won't show. Nailed to the pilaster, a backboard eliminates the need to scribe fit the fireplace's marble veneer to the pilaster profile. For economy's sake, the backboard consists of two pieces of solid oak on the exposed ends and plywood in the hidden center, all glued and biscuited together.

Frieze box has plywood upper half and solid-wood lower half. This part of the mantel top will be capped with a shelf and covered with trim, and only the solid-oak lower half will be exposed. The plywood and the solid wood are biscuited together, and blocking supports the joints.

material to; otherwise the installer would have to cut that marble to fit the profile of the pilasters. With the pilasters attached to the backboard, the installer has an easier job.

Because most of the backboard is hidden behind the pilaster (there's a ¾-in. reveal beyond the capital and base), there's no reason to make the whole backboard of solid stock. I use two pieces of 1x4 oak for the area that won't be covered by the pilasters and a plywood filler in the center. I join these three pieces with biscuits to make a 12-in. wide backboard. With the pilaster covering most of the backboard, none of the joints will be visible; the joints just have to be flush to avoid gaps between the pilaster and the backboard. I make the backboards longer than the pilasters. Once the pieces are assembled, I'll trim the backboards flush with a circular saw.

Assembling the pilasters—Now it's time to put all the pieces together. I glue up the first spacer block and center it on a capital or base so that the reveals are an even ¾ in. The pieces must be flush at the back, or else there will be gaps between them and the backboard. I nail the first block in place and position the second smaller block in the same manner. Then I attach a capital and a base to each pilaster, running screws through the backer blocks to draw the first spacer block tight (center photo, p. 91). I also nail the backboard to the pilaster so that there's a ¾-in. reveal at the capitals and bases.

The frieze is just a big box—I build the top of the mantel starting with the frieze box that will sit above the capitals and support the mantel shelf. In this case I copied an existing frieze box, though usually I rely on a scale drawing or a mock-up of the mantel to determine the frieze dimensions. This frieze box overhangs the capitals by ¾ in. both in the front and on the sides.

Although the frieze box is 15½-in. high, I use 10-in. wide solid oak only for the exposed portion of the box, the part below a 1x4 oak band, or apron. Over this apron I apply colonial doorstop and a wide crown molding, both stock items. The upper portion of the frieze box, therefore, is covered with trim, so I make the upper portion of the box from ¾-in. plywood and biscuit it to the solid oak (top photo, left).

Building the frieze box is just like building the capitals and bases in that the front and sides are mitered, glued and nailed, except I install blocks across the biscuit joints to strengthen the connections. I also nail in a couple of midspan blocks that provide backing for the bottom of the box, or the soffit, and the backing for the lid, both of which sit within the frieze box. The soffit will be exposed, so here I use oak-veneer plywood. The lid of the box can be made of anything. Its only purpose is to provide backing to secure the mantel shelf.

When installing the backing blocks, I take care to hold them 1½ in. inside the back of the box so that I can slip the box over the 1½-in. mounting ledger and have it fit snug to the wall.

On the assembled frieze box, I draw pencil lines showing the positions of the crown molding, colonial stop and 1x4 apron. I place a short

Fitting the colonial stop. The joint between the frieze box's solid-oak lower half and its plywood upper half is covered by a 1x4 oak apron to which is affixed the colonial stop.

Fastening the crown. Once the shelf is fastened to the frieze box, the crown molding is applied. The front piece goes on first, and the mitered ends are glued and nailed in place.

fingers and nail the miters together. Plywood or MDF backer blocks nailed inside the pilasters at each end support the joints (left drawing, p. 91); the backer blocks also serve as a backing both for the bases and for the capitals, which are the decorative blocks that will adorn the bottoms and the tops of the pilasters.

It's important to close up the miters as tightly as possible while the glue is still wet. I use a scrap of hardwood or my hammer handle to roll the sharp edge down just slightly, which flattens the mitered edge and closes the joint. Fine sandpaper finishes this process, leaving the mitered edges softly eased.

Making capitals and bases—The pilasters have a capital and a base; each capital and base has two 1-in. spacer blocks (right drawing, p. 91). The bases, capitals and spacer blocks are solid oak, and each one is larger than the one that it's attached to, creating a profile of ¾-in. steps.

The spacer blocks are ripped from 1x2 oak and are mitered on a chopsaw. I make them with four sides, like picture frames, with all miters glued and nailed together. I miter the corners because this mantel will be stained rather than painted, and I don't want any end grain to show.

The capitals and bases are built just like the pilasters, with a face and two sides mitered together. I use my table saw and rip fence to cut the 45° bevels, glue and nail these joints together, roll the edges and sand them. I install a lid at one end using any material handy, usually ¾-in. plywood. The lids are backing for attaching the spacer blocks, and they aren't exposed.

Backboard eliminates scribing—On this job, a marble-veneer fireplace surround will be applied after I install the mantel. To make the mason's job easier, I mount the pilasters on a backboard (bottom photo, p. 91). This board provides a straight line for the marble installer to butt the piece of crown on the frieze box with the top of the crown butted against a framing square as if it were the underside of the mantel shelf and scribe a line along the bottom of the crown. From that mark I locate and draw lines for the colonial stop and for the 1x4 apron. I cut and install the apron and the stop, mitering and gluing the outside corners (bottom left photo, facing page). The crown stays off until the shelf is installed.

Putting on the shelf—I make the mantel shelf from a single piece of 6/4 solid oak. The owner wanted the end grain to show, so I didn't miter the ends. On this job, the shelf overhangs the top of the crown molding 3 in. That overhang is consistent around the front of the mantel, too, so the width of the shelf is equal to the frieze box, plus the coverage of the crown, plus a 3-in. overhang. After ripping, crosscutting and sanding the shelf, I use a ³⁄₁₆-in. roundover bit in my router to ease the edges.

To secure the mantel shelf to the frieze box, I place the shelf upside down on my workbench and position the frieze box on the shelf, centered and flush at the back. Then I screw through the frieze lid into the mantel shelf.

Installing the crown molding—With the shelf in place, I now can install the crown molding (bottom right photo, facing page), which goes under the shelf and flush with the back of the frieze box. The main concern is getting the crown's outside miters tight.

I place the front piece of crown upside down on my compound-miter saw as if the saw table were the mantel shelf and the saw fence the frieze, and cut one end at a 45° miter. I tack this piece in place on the colonial stop, then cut the return piece. I'm no mathematician, but I am fair at trial and error, and that's the method I use to find the perfect angle on a compound-miter saw to cut crown molding. I cut scraps until they fit just right to determine the setting for the saw. Wish it were easier.

Once the saw is set, I cut the right side of the front piece first, then I hold it up to the colonial stop, where the crown eventually will be nailed. I check the fit on the right end with a short scrap, also mitered, then mark the bottom back edge of the left end. This is the ideal spot to guide in the blade of my Hitachi sliding compound-miter saw.

Once the miters are tight, I cut the backs of the return pieces so that they are flush with the back of the frieze box. Then I glue the mitered ends and securely nail all three pieces both to the stop and to the shelf.

Mantel top hangs on ledger—The mantel top is mounted on the wall at the combined heights of the pilasters, the frieze box and the shelf, plus 1½ in. of clearance for the marble flooring. I scribe this height on the wall and measure down from the mark the thickness of the mantel shelf and the frieze lid combined, which is 2¼ in. With a straightedge I scribe a line at this elevation across the wall. This line locates the top of the ledger that supports the mantel top. I rip the ledger from a piece of 2x8 and crosscut it ½ in. shy of the inside dimension of the mantel shelf,

Masking tape makes the scribe line visible. Although the pilaster's backboard eliminates the need to scribe the pilaster to the wall, the capital must be scribed to fit beneath the frieze soffit. The pilaster is shimmed up until it touches the soffit; then it's scribed and cut. Masking tape makes the scribe line easier to see on the finished capital.

allowing me a little room to adjust the final location of the mantel top. Then I fasten the ledger to the studs with 3½-in. screws and panel adhesive. On a masonry fireplace I would use plastic anchors and panel adhesive.

The frieze box slides over the ledger, and a couple of 2x4s temporarily support the box while the shelf, the soffit and the sides are scribed to the wall. The mantel top is prefinished, so I stick masking tape along the edges and scribe the lines on the tape, which makes them easy to see. I cut the top with a worm-drive circular saw. Because I cut all the scribe lines at a slight bevel, the bottom of the cut won't jut out and cause gaps to be visible between the mantel top and the wall.

Once the mantel top fits perfectly tight to the wall, I countersink holes and screw through the mantel shelf into the ledger. I use an inexpensive plug cutter by Vermont American (P. O. Box 340, Lincolnton, N. C. 28093; 704-735-7464) and make

oak plugs to cover the screw holes, then sand them smooth for the finisher to touch up.

Installing the pilasters—Because the pilasters are mounted on backboards, they only need to be scribed to the frieze soffit (photo above), not to the wall. If there were no backboard, I would install a cleat on the wall, much like the mantel-top ledger, and after scribing the pilaster to the wall, nail the pilaster to the cleat. In this case I don't need a cleat. I spread panel adhesive on the backboard and nail it right to the wall. I hold the pilasters 1½ in. above the floor, leaving plenty of room for the marble flooring to slide under the base. The resulting ¼-in. gap leaves an excuse to return to tack on some prefinished shoe molding and to take a picture (top photo, p. 90) of the finished mantel for my portfolio. □

Gary Katz is a carpenter and a writer in Reseda, California. Photos by the author.

Building a Federal-Style Mantel

Neoclassical detailing in poplar and ceramic tile

by Stephen Sewall

For the early European settlers in America, domestic life centered around the hearth. As the source of heat and the means of cooking food, it was a natural gathering place for the family. The earliest fireplaces simply had oak lintels supporting the masonry, but gradually the wooden framework around the firebox became a decorative element. By the Federal period (1790-1825), fireplaces were still the sole source of heat in the home, but weren't necessarily used for cooking. Formal fireplace mantels, with their elaborate carvings and moldings, became symbols of wealth and prosperity. The term mantel refers to the entire decorative framework around the fireplace opening, not just to the shelf above it.

I recently built a reproduction of a Federal-style mantel (photo facing page) from a house in Croton-on-Hudson, New York. My clients had spotted the mantel in a magazine article about Dutch tile and wanted to incorporate a reproduction of it in a new house they were building. Working from a photograph in the magazine, a friend of mine, David Stenstrom, made a measured drawing of the mantel, and that gave me a starting point.

My clients wanted to reproduce the entire fireplace, including a tile surround, a slate hearth and a cast-iron fireback. Therefore, the size of the firebox and the exact dimensions of the mantel were determined by the tiles we selected (Delft tiles, manufactured in Holland by Royal Makkum and imported by Country Floors, 15 E. 16th St., New York, N. Y. 10003-3104). The opening of the mantel is 12 tiles wide and 8½ tiles high. Once these dimensions were established, the slate hearth was ordered through the Bangor Slate Co. in Bangor, Me. The hearth is 16 in. wide, 72½ in. long and 2½ in. thick.

The basic frame—I built the mantel out of poplar. White pine might have been a more traditional choice, but it's more difficult to carve and dents more easily. The basic frame, or field, was 5/4 stock that I dressed to ⅞ in. (I added scribe strips later). There are two vertical pieces (side frames), each 8¼ in. wide by 58⅞ in. long, which form the base for the tapered pilasters, and a horizontal piece 14¼ in. wide by 69⅛ in. long (drawing below). I had to glue up two boards to make the horizontal piece, and when I did I oriented the growth rings the same way. When boards are glued up with the growth rings oriented in opposite directions, the resulting piece has a tendency to be wavy if it does cup.

The horizontal piece had to be as smooth as possible because I was going to rout flutes in it, and a wavy surface would mean that the flutes would be uneven. After face- and edge-

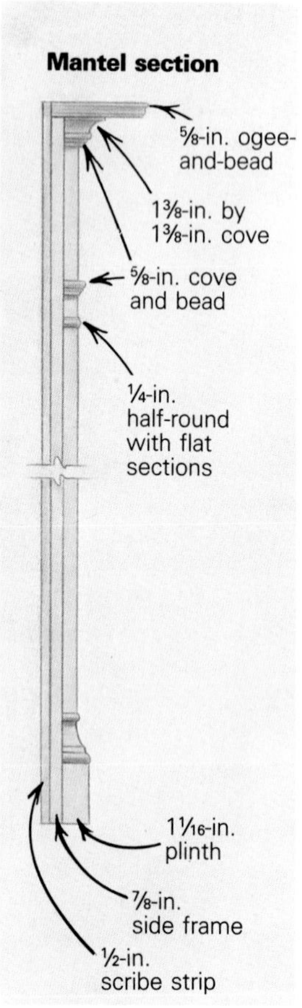

Mantel section
⅝-in. ogee-and-bead
1⅜-in. by 1⅜-in. cove
⅝-in. cove and bead
¼-in. half-round with flat sections
1¹⁄₁₆-in. plinth
⅞-in. side frame
½-in. scribe strip

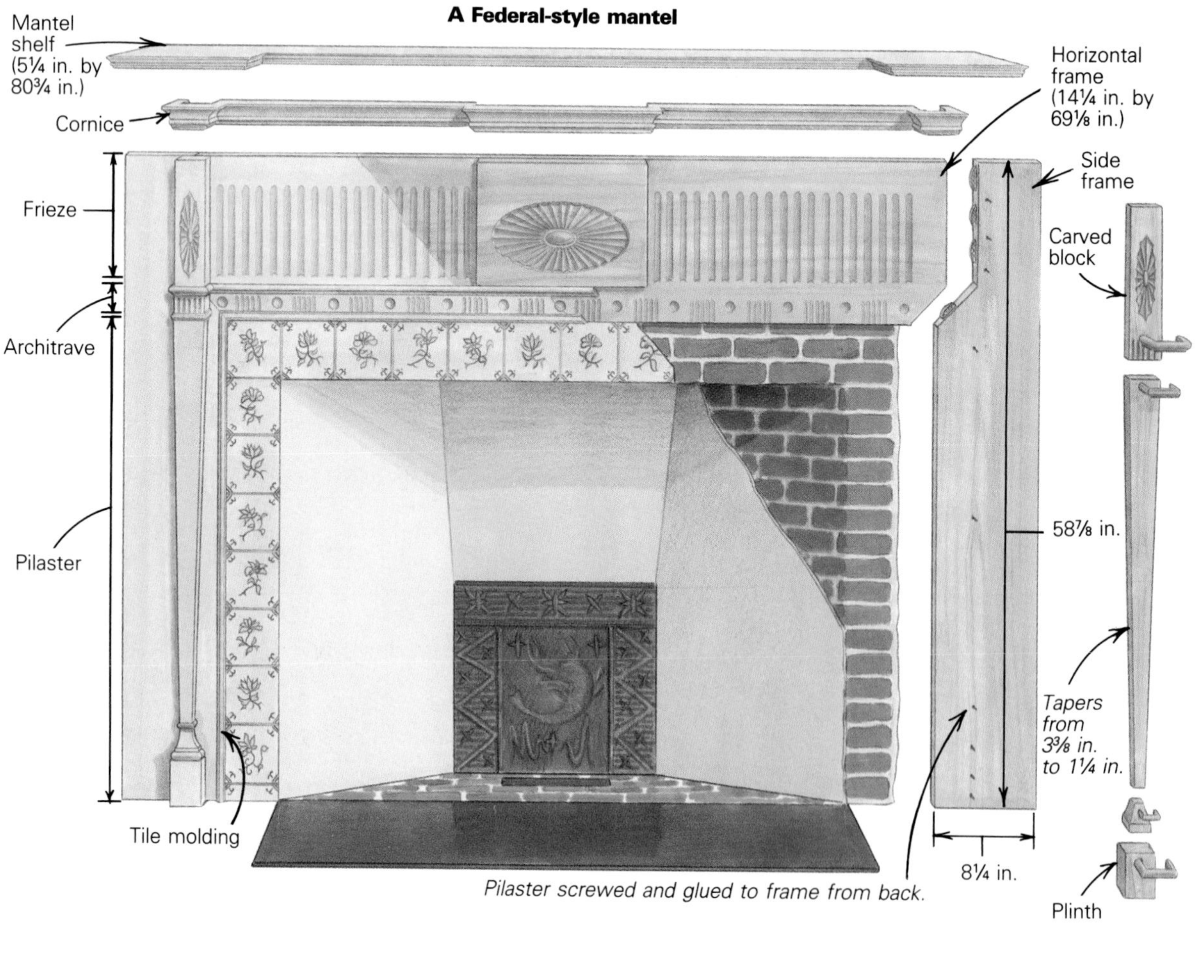

A Federal-style mantel
Mantel shelf (5¼ in. by 80¾ in.)
Cornice
Frieze
Architrave
Pilaster
Tile molding
Horizontal frame (14¼ in. by 69⅛ in.)
Side frame
Carved block
58⅞ in.
Tapers from 3⅜ in. to 1¼ in.
8¼ in.
Plinth
Pilaster screwed and glued to frame from back.

jointing the pieces, I glued the boards together, using biscuit joinery to align the pieces. Then I made one more pass to joint the entire width before planing and sanding to the final ⅞-in. thickness.

I joined the horizontal piece to the sides of the frame with a miter at the inside corner where the joint would show, then switched to a butt joint where it would be hidden behind the pilaster. I cut slots for biscuits along this joint as well, but did not glue the frame together yet because it's easier to do the routing and carving first, before gluing up.

Routing and carving flutes—The anatomy of a mantel relates directly to the elements of a classical entablature and is typically composed of two pilasters supporting the architrave, frieze and cornice, the top of which serves as the mantel shelf. I cut ⅝-in. vertical flutes down the middle of the horizontal section of the frame, which corresponds to the frieze in the entablature. Just below these larger flutes, I carved sets of smaller, 5/16-in. flutes, alternating with ⅞-in. dia., ¼-in deep holes. These smaller flutes and holes make up the architrave. I laid out all the details in pencil, drawing both sides of the smaller flutes (rather than just center lines) because they were to be carved entirely by hand. These flutes were so narrow and short that I could carve them faster than I could rout them. The bottoms of these smaller flutes run right off the edge of the board; the tile molding I applied later creates the stop.

I marked the holes between the flutes with a compass—drawing the entire circumference rather than simply marking the center point—because I was cutting them on a drill press with a Forstner bit, which has no center spur. These bits cut flat, clean-bottomed holes. The molding separating the architrave from the frieze was also penciled in, and I drew in the location of the carving block in the center of the mantel, along with the cornice molding. I laid out the ⅝-in. flutes last, keeping the area between them—known as the land—as close to ⅝ in. as spacing would permit. I marked these flutes by their centers because they would be cut with a router.

The jig for cutting the ⅝-in. flutes was made of ½-in. Baltic birch plywood with cleats screwed underneath to position the jig along the width of the board (top photo next page). The width of the hole cut in the jig is the same as the base plate of the router I used—a 3-hp Makita plunge router fitted with a ⅝-in. core box bit. The length of the hole creates the stop for the machined portion of the flute. I marked center line, top and bottom on the jig and aligned these marks with the center lines for the flutes. Because the force of the routing was resisted by the cleats underneath, I needed only one clamp to hold the jig in place. I made two passes with the router: one to remove most of the stock and a second light pass to give a clean, smooth cut.

The ends of the flutes had to be carved to make them look better. I chopped the bottom end of each flute into a convex shape with a ⅝-in. gouge. With the same gouge, I tapered

After seeing a magazine photograph of an 18th-century mantel, the owners of this house commissioned the replica shown below, including everything from the carvings to the Delft tiles, slate hearth and parged firebox.

Frame joint

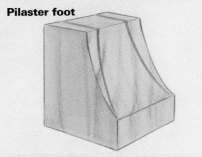

Butted portion hidden behind carved block.

Biscuits

Mitered portion remains visible.

Pilaster foot

The curves were laid out with a pattern, then cut on a bandsaw. Side cuts were made first, but weren't cut all the way through until after the front cut was made.

the top of the flute to a slight point. As modified, the flutes look to be pointing upward (bottom photo).

The pilasters—Made up of four sections, the pilasters are 58⅞ in. long overall. The top piece (1 in. by 3⅜ in. by 14¼ in.) is carved in a sunburst motif—a vertical ellipse with an oval applied to the center (more about this later). The flutes radiate out and stop in a convex chopped end. I also carved another set of 5/16-in. flutes across the bottom of the carved blocks.

The second section of the pilaster is 1 in. thick by 35¾ in. long, and it tapers from 3⅜ in. down to 1¼ in. at the foot. I cut the tapers with a tapering jig on the table saw—the same way you would cut a table leg. I started by drawing the tapered cut lines on the stock. For a jig, I found a scrap of plywood slightly longer and wider than the stock and ripped it on the table saw once, leaving the rip fence set. I lined up the cut line of the stock with the ripped edge of the plywood jig and screwed small notched blocks into the jig at either end of the stock to hold it in place. Then I ran the jig through the table saw again and had my first tapered cut. I cut one side of both pilasters, then repositioned the blocks on the jig for the last two cuts.

The foot of the pilaster was cut from a block 1 13/16 in. thick by 3⅜ in. high by 2⅞ in. wide. I made a paper pattern to lay on the block and marked the three sweeping curves. Then, on the bandsaw, I cut most of the way through two of the lines (pilaster drawing previous page). I stopped short, though, so as not to lose the third line. I cut the third line all the way through, then finished off the other cuts by eye. The plinth block at the base of the pilaster projects ¼ in. out from the foot on three sides. The block is 1 1/16 in. thick by 3⅜ in. wide by 4¾ in. high.

Assembling for humidity—When all the carving of the frame was done, I glued it up. Later, I sanded it carefully, then attached the four sections of each pilaster with glue and screws run in from the backside. At the top, where the carved sections of the pilasters covered the joint between the vertical and horizontal portions of the frame, I attached only the carved sections to the vertical boards. Their grain ran in the same direction, and the boards would therefore expand and contract in the same way. If I had attached them to the horizontal portion of the frame (the frieze) as well, the boards might have split because the grain would be running perpendicular.

By not attaching the small carving, the frieze is allowed to move freely with the seasonal changes of humidity. The mantel has been in place through two winters now, and the miter on the inside of the frame has opened up slightly, indicating that the frieze board has shrunk in width.

The sunburst carving in the center of the entablature was done on a board ½ in. thick by 11⅞ in. high by 16 in. long. I mounted this

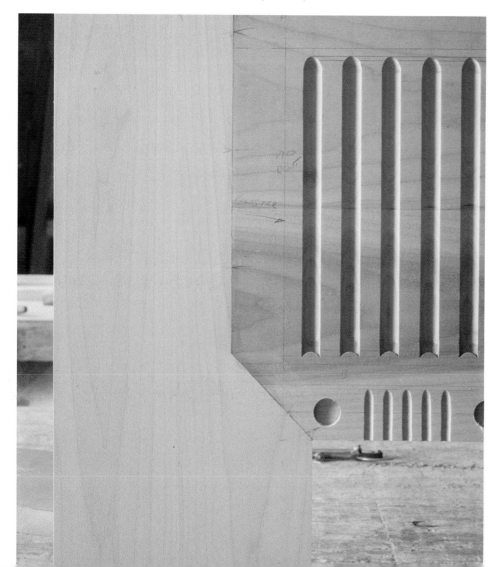

With a shop-built jig and a ⅝-in. core box bit in a plunge router, the author cut the vertical flutes that create the frieze under the mantel shelf (photo above). Where the joint would show between the horizontal and vertical members of the frame, the pieces are mitered (photo below). The butted portion will be hidden behind the pilaster. Notice how the ends of the flutes are carved so that they appear to point upward. *(Photo below by author).*

temporarily onto a piece of ¾-in. plywood and screwed a small block on the back of the plywood so I could hold the whole setup in a vise for carving. I clamped a Universal pattern-makers woodworking vise onto my regular bench vise to give me the extra height to work comfortably and prevent back strain (photo below).

To lay out the carving, I calculated the lengths of the major and minor axes (14 in. and 8 in., respectively) appropriate to the size of the block. I made up a trammel using a scrap of wood, drilling a hole in one end to hold a pencil. Two nails driven through the trammel guide it in making an ellipse. The first nail is half the length of the minor axis from the pencil, and the second nail is half the length of the major axis from the pencil. I drew intersecting lines on the block along the two axes and placed a steel square over one of the quadrants created by the intersecting lines. Using the trammel, I drew one quarter of an ellipse, keeping the nails against the square (drawing below). I repeated the procedure with each quadrant and had ellipses drawn on the block.

I penciled in the center oval by eye. Then, with dividers I laid out equal segments on the ellipse. I drew lines from these outside points to the center to represent the land between the rays. I used three sizes of gouge, each with the same sweep, to carve the rays. I had to work carefully to avoid tearing out grain. I turned an ellipsoid on the lathe, cut it in half and sanded down the section to about ⅝ in. thick. The oval was glued to the center of the carved block. The smaller carvings at the top of the pilasters were laid out the same way as the large carving in the center of the entablature. The primary difference in the carvings is that the flutes on the large carving ended with a reverse bevel, creating a flat surface to catch the light. I mounted the large carving to the frieze with glue and screws because the grain was running in the same direction.

Applied moldings — The mantel shelf is a 1-in. thick piece of poplar 80¾ in. long (the same length as the frame). It is 5¼ in. wide, allowing a ½-in. scribing space to account for an uneven wall. The front edge of the shelf jogs out 1 in. over the pilasters. I created these jogs by sawing out the center portion of the mantel on the table saw. I finished off the cut by hand and smoothed it up with a chisel.

A ⅝-in. ogee-and-bead molding was glued to the front edge of the shelf, and the shelf was set aside to be installed last. I ran a ¼-in. half-round molding around the tops of the pilasters to cover the joint between the tapered section and the carved block. This also forms a stop to the flutes on the carved blocks. I applied a ⅜-in. half-round molding to the junction of the tapered leg and the foot, then added a ½-in. half-round at the base of each foot, resting on the plinth block.

A ⅝-in. wide cove-and-bead molding that projects ⅝ in. is just under the main carving block separating the architrave from the frieze. It continues around the pilasters and dies into the frame. The same molding is used in conjunction with a large 1⅜-in. by 1⅜-in. cove to form the cornice under the mantel shelf.

All of these applied moldings were custom-made in the shop. They were glued in place where the grain direction matched that of the piece on which they were mounted. Otherwise, I just nailed the moldings, but I did glue all the miters.

I glued ½-in. scribe strips on the back of the two outside edges of the frame (no nails because I might have hit them when scribing).

The strips were 1 in. wide to allow good nailing to the wall. Before the installation, I took time to sand the carvings just enough to smooth any rough edges and didn't try to remove the tool marks. I broke all of the sharp edges of the wood so that the paint would adhere well. It was much easier to do all of the finish sanding with the mantel flat on a work bench.

The installation was easier that I had expected because the slate hearth was installed level. I only had to scribe the mantel to make it plumb to the wall, which was out some. After I nailed through the frame at the scribe strips, I slid cedar shingles behind the frieze at framing points and nailed off the mantel. There was a bow in the wall, so I did have to scribe the mantel shelf before nailing it to the cornice.

When the tile was installed — a blue and white Delft pattern called "Little Flowers" — we were careful to leave an even space between the tile and the frame. This would allow for a tile molding that would come to the edge of the tile, yet not overlap it too much. In this case the space varied from ¾ in. to ⅞ in. I made up a ⅞-in. half-round and scribed it to the tile so that it projected out from the face of the mantel by ¼ in. and nailed it to the edge of the frame.

After the tile was installed, the sides of the firebox were parged. This left a clear transition between firebox and tile and also set off the cast-iron fireback. The mantel was then painted with a traditional grey paint from Pratt & Lambert (1234 Saline, North Kansas City, Mo. 64116) called Gossamer II. The mantel required one primer coat and two finish coats. Poplar doesn't seem to cover as well as pine.□

Stephen Sewall of Sewall Associates, Inc., is a custom homebuilder in Portland, Maine.

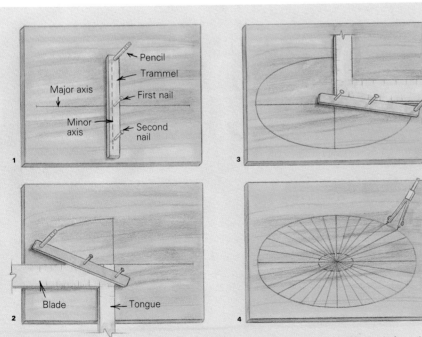

Drawing and carving an ellipse. *1) The first nail was set half the length of the minor axis from the pencil, the second nail half the length of the major axis from the pencil. 2) With a framing square over one of the quadrants, the trammel was moved so that the nails rode along the blade and tongue of the square. 3) The square was moved to each of the other three quadrants and the procedure repeated. 4) Dividers were used to step off the segments around the outside of the ellipse. The last step (photo right) was to carve the sunburst rays.*

Baronial Inglenook

Laminated mahogany arches and antique English details surround a fireplace

by Scott Wynn

The pulsating electric-blue glow of the television has long since replaced the flicker of firelight in most houses. Modern, efficient heating systems have just as surely rendered obsolete the fireplace and its messy stack of kindling and logs. Nevertheless, the comforting image of a hearth, a fireplace and the magnetic dance of the flames will probably always remain central to our concept of home. Creating a place for appreciating the fire and its tradition was one of my primary tasks in this remodeling project.

My client has a house that I think is best described as Tudoresque. It doesn't have the heavy-timbered, adzed-beam woodwork typically associated with the Tudor era, and that's fine with him. His tastes tend more toward orderly and refined trimwork—the kind that wouldn't be out of place in a ship-captain's private cabin or in a baronial library. The fireplace inglenook that I eventually designed and built not only became the focal point of the living room (photo facing page) but also set the tone for the rest of the remodel, a project that occupied my crew and me for nearly three years.

Already gutted—The job began in late spring of 1986, when my client bought a shell of a house in the hills of Berkeley, California. The former owner/builder had planned on remodeling the house, but after eight years of fitful effort, he bogged down and put the house up for sale.

We decided to stick with the existing floor plan. The existing living-room space, however, was too narrow. Widening the room over its entire length was impractical because the second and third stories were bearing on the living-room wall. We decided that a large fireplace alcove would add space as well as visually expand the room and give it a focal point.

Finding the arch—The exterior of the house was decorated in false half timbers (photo right). This seemed to make our choice of Tudor a rather obvious one, but I needed a motif to tie it all together. I took my cue from the entry to the house, which was framed by a stylized, angular Tudor arch. We settled on a softer, more elegant variation of it called a "four-center arch." Its name implies that the arcs originate at four points, but I couldn't find a formula for laying it out; I relied on my eye instead.

I used the shape of the arch liberally around the house. But by far its most notable applications were as laminated jambs and tapered casings around the fireplace and the entry-hall openings (photo next page).

Refining the casings—I made repeated line drawings on graph paper to determine the right width of the casings in relation to the size of the opening and finally settled on one 8 in. wide. I knew, however, that making the width of the casing continuous around the opening would result in an awkward and disproportionate look. Instead, I narrowed the width of the casing as it moves from the corner to the apex of the arch. I think this gives the arch a bit of tension, making it seem less static.

Somewhat more difficult to determine was the style and size of the moldings I wanted to apply to the casing. A plain casing seemed inappropriate, but it needed a feeling of depth and richness, of being "deeply carved." I settled on a half-round and a cove (drawing A, next page). To maintain the tension, these moldings had to taper to the apex of the arch along with the casing.

Scaling the arch—I began to construct these arches by blowing up the scale drawing of the curve to full size. I did this by plotting points above the springline of the arch at every foot—more frequently at the smaller radius curve—onto brown paper (drawing B, next page). I connected the points, sketching a curve and correcting it as needed. Once I was satisfied with the curve, I drew a second line ⅞ in. outside of it. This line compensates for the five layers of ⅛-in. thick laminae of the arched jamb, and the two ⅛-in. thick pieces of plywood that would be applied to the bending form for a clamping surface (drawing C, p. 101). Using carbon paper, I transferred this line to a sheet of ⅝-in. MDF. Note that the form is concave—not convex. That's because I like to have the finished side where I can see it during glue-up, which gives me a little bit more

Evolving arches. **The author used the arch over the front door (photo below) as a point of departure for a unifying design element. Inside, the inglenook (photo facing page) is framed by a softer, curved version of the arch made of laminated mahogany. The concrete fireplace surround repeats the same arch. The surround blocks were cast on site in melamine-coated particleboard forms held together with screws for easy disassembly.**

control on these critical laminae should they need shifting or an extra bit of pressure.

I used MDF for the form because it's cheaper than plywood and stronger than particleboard. After cutting out the basic shape, I used a rasp to take out any inaccuracies, checking constantly with a square to make sure the edges remained perpendicular to the face of the form. When I was satisfied with the line, I took a second piece of MDF and cut it to within ⅛ in. of the line, then clamped it to the first and trimmed it to the exact profile with a flush-trimming bit in a router. Accuracy is critical here because any variation between the two sides of the form will cause the laminae to shift out of alignment as they are glued up. I ripped the scrap into strips to use as spacers between the two sides of the form.

Next I glued and nailed two layers of ⅛-in. bending poplar to the business edge of the form to provide a smooth and consistent surface to back the clamped laminae. I took care to set the nails so that no dimples would mar the clamping surface. Then I drilled 1½-in. dia. holes near the form edge at roughly 4 in. to 8 in. apart to allow room for the clamps (drawing C, p. 101).

Finally I could begin assembling the arches; I started by resawing mahogany for the jambs. I chose a piece of 1x8 for the faces that had a striped figure, and set aside the first two pieces that came off the saw so I could later bookmatch them. The faces are backed up by one more layer of mahogany, and three of poplar bending plywood. I oriented the face grain of two of the layers of bending plywood so they were perpendicular to the length of the arch; this would counteract springback from the mahogany. I've found that if you orient the face grain of all the layers of bending poplar perpendicular to the curve, the radius will actually decrease as the glue dries.

I also cut enough of the bending plywood to use as clamping cauls for the curved portion of the arch. For the cauls, the grain runs parallel to the curve to help spread the clamping pressure. On the straight parts of the arch, I used a piece of scrap MDF as a caul. All the cauls were about a ½ in. wider than the jamb to ensure full bearing.

Glue-up—I used Titebond, an aliphatic resin glue (Franklin International, Corporate Center, 2020 Bruck St., Columbus, Ohio 43207; 614-443-0241), to bond the laminae together, and spread it around with a 4-in. putty knife. I find this method to be faster than using a paint roller, though it takes practice to get the glue thickness consistent. Either way, it's a messy job. To make sure I could later release the jamb from the form, I lined it with wax paper.

I collated the glue-covered laminae on a flat workspace, clamped one end to keep them from shifting around, and then took them to the form. Jamb in place, I started at the smallest radius and worked outward, setting the clamps as I went.

Once the glue set (I allowed at least 12 hours) I removed the jamb, scraped off the excess glue along the edges (this speeds curing) and set the jamb aside while I glued up the next one. Before I started to joint any of the jambs, I let them sit for at least 24 hours after I removed them from the form.

Straight edges—To clean up the edges of each jamb I removed the remaining lumps of glue with an old iron plane, sighting along the edges to get them as straight as possible. Normally, I put a laminated jamb on the jointer, and with a helper, maneuver it over the knives, taking off just enough to get a straight line.

That's the typical sequence, but in this case, the long pieces wouldn't fit on the jointer in our small shop; therefore, I ran them through the table saw, making sure to keep the back of the jamb tight to the table as it passed the blade. A long fence to serve as a reference plane is practically a necessity for this operation.

After taking a light pass on one side, I used a hand plane, when necessary, to take out imperfections. Once I was satisfied with the edge, I ripped the opposite edge, leaving enough along the first joined side to be able to rip a clean edge and still get my finished width.

Casing layout—The casings for each arch section join at the tight radius in the corner. I plotted their angle and orientation by doing a full-size template on brown paper (drawing D, facing page). I joined them with biscuit wafers at the miter, and then glued the casings to the jambs. Once the glue was dry I used a saber-saw to cut away the larger portion of the waste. I followed this by using a router with a flush trimming bit to match the casing to the exact curve of the arch. Then I cut the cove in the edge of the casing with a piloted cove bit.

Next, I used a straight bit with an adjustable pilot to cut a slight rabbet at the edge of the cove. This makes a narrow reveal that mimics the setback of a casing around a door. I did this on all the arch pieces, and then, using my

Arched jambs. Laminated mahogany jambs trimmed with tapered casings frame the entries to the dining room, living room and the inglenook in the distance. The arch also turns up in smaller scale as a border atop the fret-sawn panel applied to the newel post in the center of the photo.

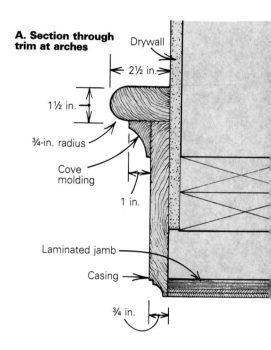

A. Section through trim at arches

Drywall
2½ in.
1½ in.
¾-in. radius
Cove molding
1 in.
Laminated jamb
Casing
¾ in.

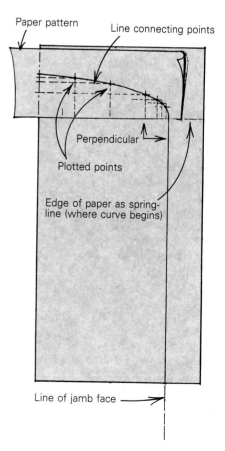

B. Laying out the arch

Paper pattern
Line connecting points
Perpendicular
Plotted points
Edge of paper as spring-line (where curve begins)
Line of jamb face

paper template, I laid out the line of the outside edge of the arch on the casing and cut it with a sabersaw.

Coves, half-rounds, donuts—I fashioned the moldings with a router, using standard cove and roundover bits. The half-round was done with passes on opposite sides of a mahogany 2x3. Even though the bit has a pilot, I used a fence to guide the router. That's because on the second pass, the pilot rides on the previously cut radius, preventing it from cutting a true half-round. I glued and biscuit-joined these moldings to the straight portion of the jamb, running them slightly long.

To do the radius portions of the half-round moldings I glued four blocks of mahogany together, with the grain oriented diagonally, and attached them to a piece of plywood (drawing E, below). I mounted this assembly on the faceplate of my lathe, turned the blocks into "donuts," and then ran the router over them with the half-round bit. Then I bandsawed them off the plywood backing and glued them in place, trimming moldings on the straight section of casing as required to match.

The radiused cove sections were too small to make easily with a router, so I bandsawed a piece to fit and installed it unshaped. After the adjoining moldings were installed, I carved,

scraped and sanded the little blocks to match.

The moldings, as they go from the radiused corner to the apex of the arch, are not only tapered but curved. I roughed out the moldings on the bandsaw, then closed in on the final profiles with router bits. I finished them with molding planes. On the cove portion, I actually took a piece of the standard cove molding and bandsawed a bit off both sides to get the taper. Then I finished the cove with a molding plane, flexed it into place and glued it.

Jambs on site—What I took to the site looked like a truckload of huge hardwood boomerangs. I had sections for two arches, each one

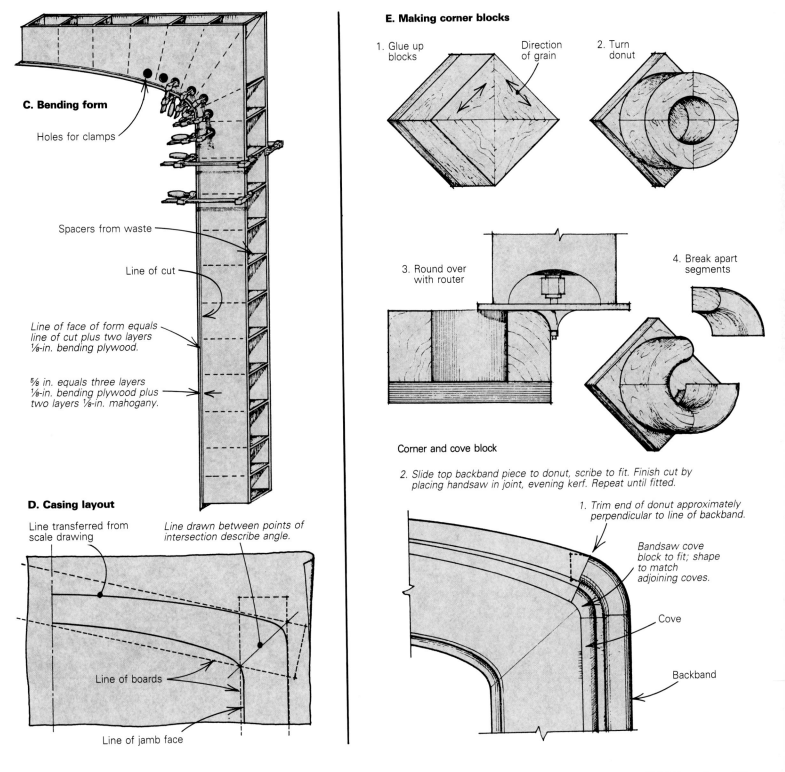

C. Bending form

Holes for clamps

Spacers from waste

Line of cut

Line of face of form equals line of cut plus two layers ⅛-in. bending plywood.

⅝ in. equals three layers ⅛-in. bending plywood plus two layers ⅛-in. mahogany.

D. Casing layout

Line transferred from scale drawing

Line drawn between points of intersection describe angle.

Line of boards

Line of jamb face

E. Making corner blocks

1. Glue up blocks   Direction of grain   2. Turn donut

3. Round over with router

4. Break apart segments

Corner and cove block

2. Slide top backband piece to donut, scribe to fit. Finish cut by placing handsaw in joint, evening kerf. Repeat until fitted.

1. Trim end of donut approximately perpendicular to line of backband.

Bandsaw cove block to fit; shape to match adjoining coves.

Cove

Backband

divided into finished halves with moldings mounted on both sides. I made the sections so that I had about 3 inches of play per side between the back of the jambs and the rough framing.

To fit the sections, I first checked how level the floor was across the opening to find the low side. Then I slid the jamb assembly for that side over the wall, plumbed it, and scribed it to the floor. I marked a cutline at the apex, removed the first assembly, slid the second assembly over its wall and repeated the procedure. I trimmed the arch tops to the prescribed marks, and then slid them back in place. To get the fit just right at the top, I butted the two halves together and ran a Japanese draw saw, made for cutting hardwoods, between the butted ends to create an equal gap. Then I pushed them together for a tight fit. Last, I nailed the assemblies to the wall. They are secured to the framing by toenailed finish nails at the base and finish nails run through the casings into the studs.

Surrounding the fireplace—The richness of the wood and the complexity of the trim in the rest of the house demanded that the fireplace detailing be of great intensity, if it was to be the focal point of the house. To inform my designs, I researched English residential architecture, with an emphasis on the 16th

and 17th centuries. One excellent reference I used is *A History of the English House* by Nathaniel Lloyd (Omega Press Ltd., 1 West St., Ware, Hertfordshire, England). From this reference I learned about the individual decorative elements of the fireplace and found useful examples that I could mix and match to make an elegant fireplace surround for this job.

To get my proportions, I began with the firebox opening. We wanted this to be as large as we could get it, limited only by the ratio of the opening to the cross section of the flue. We decided to use a manufactured metal flue, as the cost of a seismically reinforced masonry flue would be prohibitive. Our maximum flue area was 154 sq. in. (14-in. flue). Working on the theory that says the ratio of firebox-opening-to-flue area should be 10 to 1, we could have a firebox opening of approximately 38 in. by 40 in.

The fireplace opening carries the motif of the arch. By code, any wood has to be at least 12 inches away from the opening. I laid out these dimensions on a piece of paper, along with the height and width of the alcove. Then I divided the surround into individual zones (drawing below). Pilasters flank the opening, supporting a mantel. Above that is an overmantel, which is capped at the ceiling by a crown molding.

I decided that a single large mantel would be too rustic. Instead, I combined a more sophisticated curved mantel that reiterated the crown molding at the ceiling. The overmantel between them would be divided into horizontal sections: a base, a top frieze and a band between them containing three decorative panels (drawing below). These would be divided by pilasters similar to the ones that support the mantel.

Refining the zones—Having arrived at the proportions of the different elements, I began to articulate them. In some places, I used moldings to differentiate or highlight areas. For example, a watertable molding with a cove accents the step between the base of the overmantel and its central panels, and an astragal heightens the border between the frieze board and the carved panels.

The overmantel pilasters (photo facing page) are Jacobean in their design: a flat-faced column tapering in two dimensions to the plinth. The plinth has a double-tapering cove top, and contrary to tradition, the capital tapers to a waist. The applied geometric patterns, such as the diamond plaques, are typically Jacobean as well. I chose the sections of the moldings and their proportions based on photos of work from that era. I also used

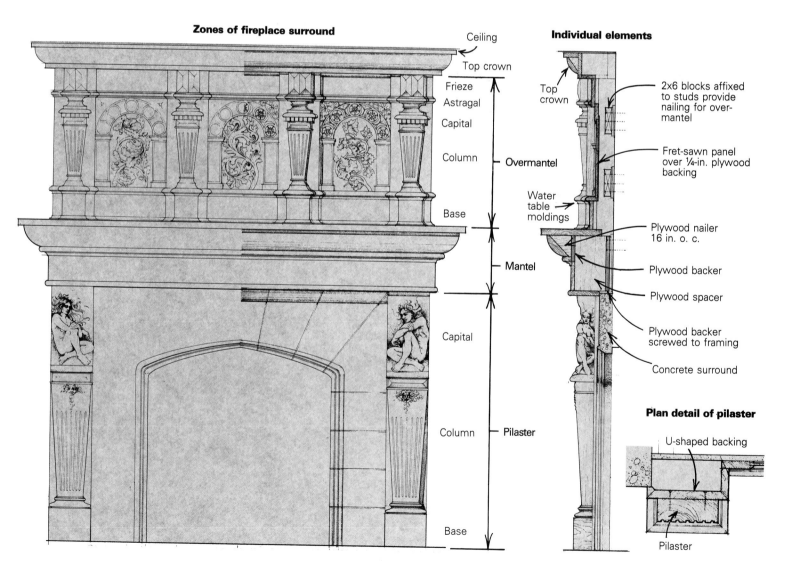

Zones of fireplace surround

Ceiling
Top crown
Frieze
Astragal
Capital
Column — Overmantel
Base
— Mantel
Capital
Column — Pilaster
Base

Individual elements

Top crown

2x6 blocks affixed to studs provide nailing for overmantel

Fret-sawn panel over ¼-in. plywood backing

Water table moldings

Plywood nailer 16 in. o. c.

Plywood backer

Plywood spacer

Plywood backer screwed to framing

Concrete surround

Plan detail of pilaster

U-shaped backing

Pilaster

some dentils, which seem to be common to nearly all eras.

The most difficult part, however, I had yet to tackle. Namely, how much projection does everything have? It was much harder for me to visualize the relative depths of the parts—how much they would need, how much would be too much—than perhaps any other aspect of the design.

I began by determining priorities based on visual importance. The mantel would be the strongest line, the crown at the ceiling would step back from that but still make a strong cap at the ceiling. The pilasters needed to be unmistakably unified with the mantel, while the overmantel had to fall somewhere between the mantel and the wall.

To assign dimensions, I started with the pilasters. I didn't want a glue line to show, so I made them from 16/4 stock. They were applied over backing pieces that are U-shaped in plan (detail drawing, facing page). Their combined thickness, plus the curve of the mantel's crown molding, determined the projection of the mantel. Simple enough. The overmantel was less so. I felt its deepest part should still be slightly proud of the wall, so I began stepping forward from there with the different elements, in ¾-in. and 1½-in. increments. I adjusted the steps so that the foremost part, the

overmantel pilasters, were still about even with the major pilasters under the mantel. The top crown forms kind of a soffit over the whole thing (drawing facing page).

Making the parts—I took care to avoid fixed crossgrain construction—instead, large panels float inside their frames. I also used as few nails as possible, as they are always visible. The mantel is backed by a piece of plywood for a nailing surface. The overmantel is a frame-and-panel assembly joined with biscuits. Its pilasters are screwed to it from the back.

I cut the pilasters from solid stock and incised their converging flutes with a router and a corebox bit. I did this by carefully plotting the location of the flutes on the finished face before cutting the side tapers. Then I ripped a taper parallel to the center flute and used it as a guide for my router to cut the flute. Next I ripped tapers parallel to the next set of flutes and again used them to guide my router. I did this successively until all the flutes were cut, then ripped the final shape of the pilaster.

Because I needed less than 15 ft. of most molding shapes, I made them in the shop using router bits, or roughly shaped them with the table saw and then worked them down with planes. Any flat spots left by this step I removed by using either flexible scrapers or scrapers ground to the exact profile. The latter I ground from throwaway Japanese sawblades. I find they remove stock very quickly, and frequently there's no need for follow-up sanding.

Carved panels and capitals—For decorative motifs for the overmantel panels I referred to two sources: *Historic Ornament: a Pictorial Archive*, by C. B. Griesbach, (Dover Publications, 31 E. 2nd St., Mineola, N. Y. 11501, 1975. $10.95, softcover, 280 pp.; 516-294-7000) and *Manual of Traditional Woodcarving*, by Paul N. Hasluck, (Dover Publications, 1977. $10.95, softcover, 568 pp.). I found some appropriate floral decorations from the era and redesigned them to fit our needs. Incidentally, I found a lot of other details in these two books for miscellaneous decorations that worked into this job.

I used a technique called "fretted carving" to make the panels. This is simply a jigsawed design applied to a panel before the final details are carved. While the finished panels look relieved (photo left), this method is a lot faster.

I had originally envisioned the pilasters topped by crouching medieval figures—gargoyles of some sort. These evolved, however, into figures of repose and meditation more classical and refined (photo, p. 98) than the figures I first imagined. These were my concession to the present: I didn't really set out to make a reproduction of a style but rather something more of an impression of a style.

Before I picked up the chisels, I made clay mockups of the figures to help me visualize them. Then I sketched their outlines on the faces and sides of the 16/4 pilaster stock and cut away the major waste with a bow saw.

Next I used chisels and gouges to distinguish the levels of different elements. Once these were established I refined the volumes and details of the figures, taking care to not let some areas get too far ahead of others.

I don't use sandpaper unless absolutely necessary, so to finish the figures I made burnishing cuts and used tiny curved files called "rifflers." The trick to burnishing is to make cuts that are nearly parallel to the surface of the work. Opposing cuts must meet exactly so there is no torn wood where a chip has been "pried" out. The rifflers make it easy to clean the crevices.

Installing the mantel—The alcove's paneling had already been installed, so I took the mantel to the job site in four pieces (mantel, overmantel, and two pilasters). Each piece had been prefinished in the shop, with three coats of sanding sealer and four coats of lacquer. I also brought enough crown molding to finish the alcove and enough horizontal banding to extend the lines of the mantel and the base of the overmantel into the corners. I coped one end of all the moldings in the shop—the large crown molding is particularly difficult to cope, so doing it in the shop on the bandsaw was much easier than field-fitting.

The day before we installed the mantel, I secured horizontal 2x6 blocks to the wall behind the overmantel. I positioned them exactly so that I could nail through the sides of the overmantel, out of sight, into the end grain of the 2x6s. There was nothing to nail to across the center of the mantel—so it had to be nailed near its ends.

I affixed the mantel's wide shelf to the bottom of the overmantel assembly. That allowed me to fit the many pieces of narrow cove molding that wrap around the bases of the small pilasters while still in the shop. Also, leaving the wide shelf off the mantel assembly gave me access to the mantel's plywood backing from above. Once I had it positioned atop the pilasters, I affixed the mantel to the studs with nails through the plywood (drawing facing page). To attach the two mantel sections, I toenailed the crown molding to the bottom of the wide shelf.

Truth be told, the only significant hitch that we encountered during installation was a result of my not accurately measuring the space for the overmantel. The embarrassing thing is, I thought we had enough room for a ½-in. reveal all around the top, but because of variations in the framing and taping (which I thought I had accounted for) the overmantel was about ⅛ in. too tall. Because of its assembled weight, this was no small obstacle. Luckily it was hitting the ceiling in the back, out of sight, so by removing wood there we were able to get it in, though not without lifting it up, driving it on and prying it off three times.□

Scott Wynn is an architect/contractor who also designs and builds furniture in San Francisco. All drawings are by the author. Photos by Charles Miller.

Pilasters and fretwork

Wynn used a router and core-box bit to cut the converging flutes in the pilasters. He started with a piece of stock with parallel sides and used one side to guide his router fence as he cut the center flute. Then he tapered the sides in successive steps, each time using them to guide the router fence. Wynn began his fretwork panels by first halving a panel on the bandsaw. Then he rough-cut the floral pattern from one of the halves and glued it back onto its mate before carving, ensuring a perfect match in color and grain.

A Built-In Hardwood Hutch

When working with solid wood, joinery techniques must accommodate seasonal movement

by Stephen Winchester

The opening. New studs frame the sides, but the back wall of this former closet was straightened with 1x3s and shims.

Ash matches. A new ash hutch built into an old closet looks like the chestnut woodwork of the original room. The hutch was finished with two coats of Minwax Polyshades—half maple and half walnut—followed by a slightly thinned top coat.

I love an old house. Working on one makes me appreciate the skill of the carpenters who came before me. It's amazing to see the level of craftsmanship the old-timers attained using only hand tools—especially in their trimwork. I recently renovated an early 1800s farmhouse in New Hampshire that had some beautiful chestnut trim. I got the chance to match this woodwork when I added a family room with a built-in hutch (right photo, above).

I made the new family room by removing a wall between two small rooms. There was a clos-et in each room, one on both sides of the wall, and when the wall came down, the closet area was a natural location for the built-in hutch. Built-ins ought to look good and last a long time, so this hutch was built of solid hardwood and designed to accommodate wood movement (drawings facing page). But before I started building, I straightened and leveled the closet area.

Roughing in the hutch—New studs on the left and right made the sidewalls plumb and straight, but there wasn't room on the back wall for new studs. So I straightened the back wall with shims and 1x3 strapping (left photo, above). At the bottom I tacked a 1x3 across the old wall and into the old studs. Placing a straightedge on the 1x3, I tucked some shims behind the low spots to bring them out to the straightedge. Next I tacked a 1x3 to the top, again shimming it straight. Then I tacked on more horizontal 1x3s 16 in. o. c. Moving from left to right, I held the straightedge vertically, against the top and bottom strapping, and shimmed the intermediate strapping out to the straightedge. The wall was

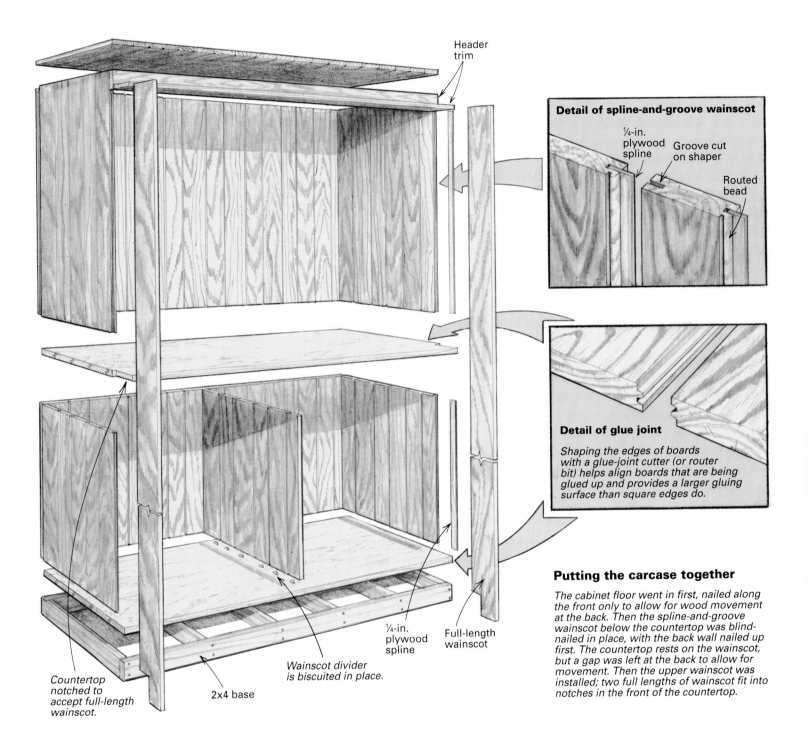

Header trim

Detail of spline-and-groove wainscot

¼-in. plywood spline

Groove cut on shaper

Routed bead

Detail of glue joint

Shaping the edges of boards with a glue-joint cutter (or router bit) helps align boards that are being glued up and provides a larger gluing surface than square edges do.

Putting the carcase together

The cabinet floor went in first, nailed along the front only to allow for wood movement at the back. Then the spline-and-groove wainscot below the countertop was blind-nailed in place, with the back wall nailed up first. The countertop rests on the wainscot, but a gap was left at the back to allow for movement. Then the upper wainscot was installed; two full lengths of wainscot fit into notches in the front of the countertop.

Countertop notched to accept full-length wainscot.

2x4 base

Wainscot divider is biscuited in place.

¼-in. plywood spline

Full-length wainscot

straight when all the pieces of strapping were even with each other.

The hutch rests on a 2x4 base; I installed it level by shimming the low end and nailing it to the new 2x4 walls on each side. With the new level base, I didn't have to scribe the cabinet sides and back to the floor, which had a big hump in it.

Chestnut substitute—Chestnut was once used for almost everything in a house, from sheathing to door and window frames to trim. But during the first part of this century, a blight wiped out almost every American chestnut tree. Today, you can get salvaged chestnut from old buildings or get it resawn from beams or sheathing, but it's expensive. I chose white ash instead, which has about the same grain pattern and texture as the chestnut woodwork on this job. But ash is hard, so it's more difficult to work than chestnut.

Gluing up wide boards—The cabinet floor and the counter were glued up out of several boards, as were the wide shelves for the bottom cabinet. To joint and join the boards in one step, I used a glue-joint cutter in the shaper. (Jointing is the process of straightening a board's edge or face and is typically done with a jointing plane or with an electronic jointer. Joining is the process of connecting two boards.) The glue-joint cutter makes edges that look something like shallow finger joints (detail drawing above). These edges align the boards and provide a larger gluing surface than simple square edges do. Glue-joint bits are also available for use in router tables.

First I lined up the boards so that their grain matched, and I marked them so that they

Drawings: Bob Goodfellow

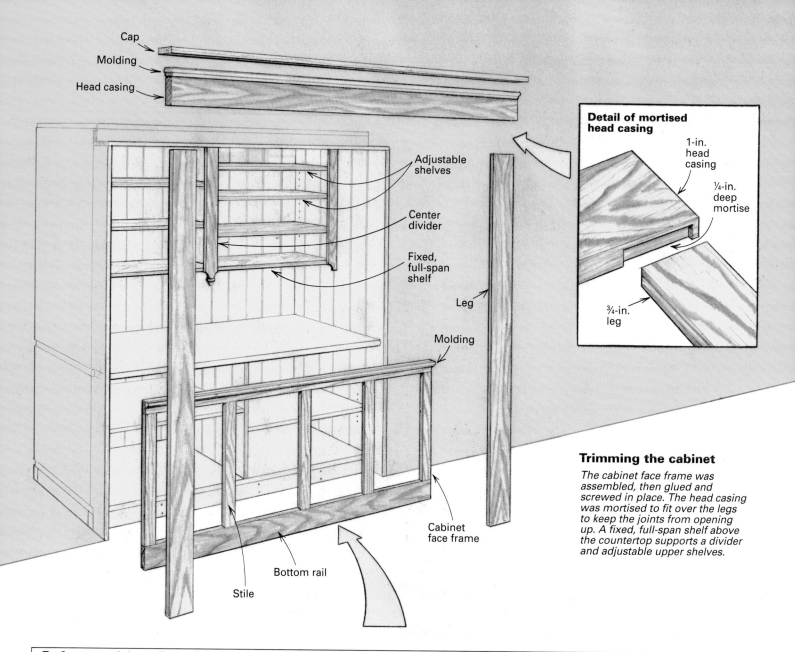

Cap

Molding

Head casing

Adjustable shelves

Center divider

Fixed, full-span shelf

Leg

Molding

Cabinet face frame

Bottom rail

Stile

Detail of mortised head casing

1-in. head casing

¼-in. deep mortise

¾-in. leg

Trimming the cabinet

The cabinet face frame was assembled, then glued and screwed in place. The head casing was mortised to fit over the legs to keep the joints from opening up. A fixed, full-span shelf above the countertop supports a divider and adjustable upper shelves.

Pocket-screw joinery. **To attach the bottom rail to the stiles, a spade bit makes a pocket hole that's 1½ in. short of the rail's edge (left). A pilot hole is then drilled up through the edge to connect with the pocket hole (middle), and the boards are glued and screwed together (right).**

wouldn't get mixed up during the glue-jointing operation. I used numbers—1s on the first two adjoining edges, 2s on the next two and so on.

I don't have a wide planer, so I had to flatten the glued-up boards with a belt sander. With a 60-grit belt, I sanded across the grain first, then with the grain. Then I used a 100-grit belt and finished with a 120-grit belt. The countertop, the most visible of these wide boards, was finished using 180-grit paper on a random-orbit sander.

Spline-and-groove wainscot—One of the original small rooms had beaded wainscot all the way around, so I decided to use beaded wainscot inside the hutch. To make the wainscot, I ripped ash boards on the table saw into random widths, from 5¼ in. to 3¼ in.

To join the pieces, I used a spline-and-groove joint rather than a tongue-and-groove joint (detail drawing, p. 105). First I jointed the edges of each board. To make the groove, I used a ¼-in. straight

cutter on the shaper, but a ¼-in. slotting cutter in a hand-held router or a dado-blade assembly in the table saw would work, too. I centered the ½-in. deep groove on the edge of the board. The ¹⁵⁄₁₆-in. splines were ripped from ¼-in. plywood. I didn't use biscuit joinery because, when wainscot shrinks, gaps appear between the biscuits. A full spline looks like a solid tongue.

Using a beading bit, I beaded one edge of each board to match the original chestnut wainscot.

Installing the wainscot—I installed the floor of the cabinet first, flush to the front of the 2x4 base. I nailed the floor at the front only and left a ⅜-in. space at the back to allow for wood expansion.

I put up the wainscot for the bottom half of the hutch by blind-nailing through the splines and into the walls as I would any T&G material. I didn't glue the splines because each piece of wainscot should expand and contract independently. This wainscot rests directly on the cabinet floor; if the floor butted into the wainscot, a seam would open. I avoided visible seams in the corners by putting up the back wall first and then butting the sidewalls into it. And I allowed for expansion by installing the first board on the back wall ⅜ in. from the corner.

I also made a wainscot divider for the bottom cabinet. It was biscuited to both the floor and the underside of the counter. I used just a dab of glue in each biscuit slot to prevent any unnecessary glue squeeze-out.

The counter sits on the wainscot. Before I installed the counter, I notched its two front edges, which would allow an entire length of wainscot at the front of each sidewall. Along the sides, the counter is nailed into the wainscot so that it stays put, but to allow for expansion and contraction, the back edge of the counter isn't nailed.

Now I was ready to put the wainscot in the top of the hutch. I set the wainscot on the counter and blind-nailed it through the splines into the walls. Putting the wainscot up in two sections, bottom and top, eliminated the wood-shrinkage gaps that would have resulted from running the wainscot from floor to ceiling and butting the counter into the wainscot. With the front edges of the counter notched, I installed the front pieces of wainscot on each sidewall. Because the unit is recessed into the opening, I wanted a full length of wainscot from floor to header with no seam.

Pocket-screw joinery—In my shop, stiles and rails for the face frame were cut to width but not to length. Stiles and rails are the vertical and horizontal frame pieces, respectively.

I assembled the face frames on site. I cut the stiles and the rails to length and clamped them to the cabinet to check the fit. After some slight trimming on a compound-miter saw perfected the face-frame joints, I laid the stiles and the rails on the bench and screwed them together.

The top rail was narrow enough to allow the stiles to be joined to it with screws driven straight through the edge. But the bottom rail of the cabinet was wide, and the intermediate stiles butted into it, so here I screwed the rail to the stiles through pocket holes. A pocket hole is a cut made on the face of a board that doesn't reach the board's edge.

There are several jigs on the market to make pocket holes—from simple guides for a hand-held drill to dedicated pocket-hole machines. I don't have any of them, so to make the pocket holes to assemble this face frame, I used a spade bit, starting the hole with the drill held vertically and tipping the drill back as I fed the drill bit in (left photo, facing page). The pocket hole ended at a

Matching molding

A table saw and a shaper were used to make ash molding (right) that matches the original chestnut trim (left).

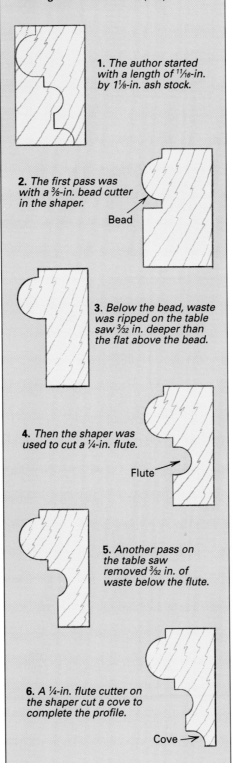

1. The author started with a length of ¹¹⁄₁₆-in. by 1⅛-in. ash stock.

2. The first pass was with a ⅜-in. bead cutter in the shaper.

Bead

3. Below the bead, waste was ripped on the table saw ³⁄₃₂ in. deeper than the flat above the bead.

4. Then the shaper was used to cut a ¼-in. flute.

Flute

5. Another pass on the table saw removed ³⁄₃₂ in. of waste below the flute.

6. A ¼-in. flute cutter on the shaper cut a cove to complete the profile.

Cove →

mark 1½ in. from the edge of the rail. Then I drilled a pilot hole in the edge of the rail at an angle up through the pocket hole (middle photo, facing page). Finally, I squeezed a generous amount of glue between the stiles and the rails, clamped them together and ran the screws in (right photo, facing page).

After the glue was dry, I sanded the joints flush and installed the face frame. I glued the bottom rail to the front edge of the cabinet floor and screwed the top rail to the underside of the countertop (drawing facing page).

Mortised head casing—The trim, or casing, around the hutch was installed next. The ¾-in. thick side pieces, or legs, went on first; I ran them ¼ in. long at the top. The 1-in. thick top piece, or head, was mortised to fit over the legs (detail drawing facing page). You could think of this as being a mortise-and-tenon joint, with the legs being the tenons. I set the head on top of the legs and with a sharp pencil traced the outline of each leg onto the bottom edge of the head. Then I scored the marks with a sharp knife. Scoring makes for a cleaner mortise. I mortised these sections of the head a good ¼ in. deep with a hinge-mortising bit in my small router. Finally, I used a chisel to square the corners of the mortise. This joint practically guarantees lasting beauty: If the header shrinks, the joint still looks tight.

I wanted the molding under the front of the counter and at the top of the head casing to match the original molding at the top of the doors and the windows (photo above). This molding wasn't something I could have picked up at the lumberyard, and I couldn't find any cutters the right shape, so I combined two different shaper cutters to make the molding (drawing left). The result was a perfect match.

Making doors—I made frame-and-panel doors for the cabinet at the bottom of the hutch.

The door stiles are 2 in., the top rail is 2½ in., and the bottom rail is 3½ in. After cutting the pieces to size, I used my shaper to mold the inside edges of the frame, cut the panel groove and make the cope-and-stick joint between the stiles and the rails.

I assembled the frames dry to check the door size and to get the panel size. I allowed ⅛ in. on each side of the panel for expansion. The ash panels on these doors were raised (beveled around the edges) on the shaper, so I glued up the boards with square joints to make the wide panels. If I had used the glue-joint cutter, the glue-joint profile would have been visible when I shaped the raised edges.

To be sure everything fit, I dry fit the panel within the frame before gluing up. Then I glued the doors and clamped them. I used a small amount of glue on the joints because the squeeze-out could glue the panel in place, and the panel should be free to expand and contract. □

Stephen Winchester is a carpenter and woodworker in Gilmanton, N. H. Photos by Rich Ziegner except where noted.

Simple Closet Wardrobe

A biscuit joiner and a layout jig make construction quick and easy

by Jim Tolpin

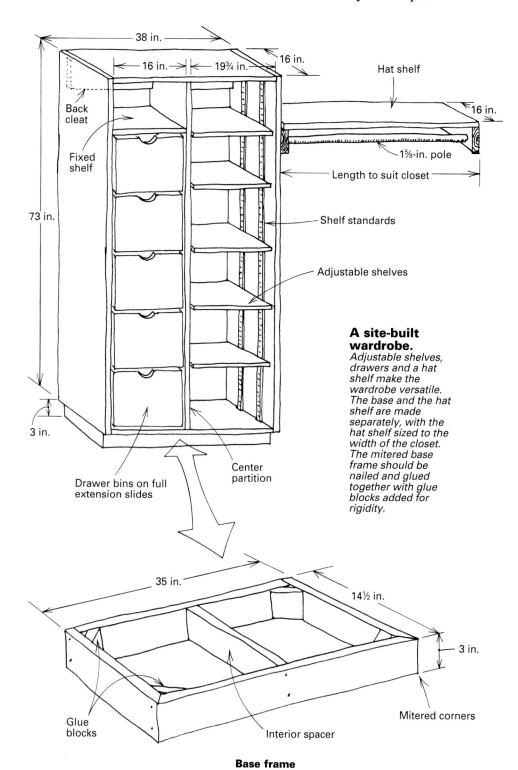

Hat shelf

16 in.

1⅝-in. pole

Length to suit closet

Shelf standards

Adjustable shelves

38 in.

16 in.

19¾ in.

16 in.

Back cleat

Fixed shelf

73 in.

3 in.

Drawer bins on full extension slides

Center partition

A site-built wardrobe.
Adjustable shelves, drawers and a hat shelf make the wardrobe versatile. The base and the hat shelf are made separately, with the hat shelf sized to the width of the closet. The mitered base frame should be nailed and glued together with glue blocks added for rigidity.

35 in.

14½ in.

3 in.

Glue blocks

Interior spacer

Mitered corners

Base frame

Closet storage systems are in vogue these days. Homeowners are buying prefabricated drawer units and paying closet specialists to install expensive epoxy-coated wire bins and shelves. Yet a handsome closet wardrobe can be constructed in just about a day by anyone who learns how to use a biscuit joiner and a shop-made layout jig.

A simple layout technique makes the process foolproof. The same approach is also effective for many other kinds of case pieces like bookshelves and cabinets. Plus, all of the work can be done on the job site with just a few tools and a workbench no more elaborate than a flat surface on a pair of sawhorses.

This surprisingly compact wardrobe unit (photo facing page) features a bank of slide-out sweater drawers, adjustable shelves and a hanging locker with a hat shelf. A typical bedroom closet is 2 ft. deep and 5 ft. or 6 ft. wide, so I make my wardrobes 16 in. deep and devote half the width of the closet to the drawers and the shelves. The size of the accompanying hat shelf and closet pole can be adjusted to fill the rest of the space.

The project can be built from either ¾-in. boards or ¾-in. sheet stock like hardwood plywood or melamine (particleboard covered with a thin plastic laminate). Some suppliers stock melamine that is precut to 16-in. widths with a thin strip of the same laminate applied along one edge. This gives the piece a finished look.

Making the base frame—The wardrobe is essentially a tall box with a vertical divider down the center. One side is devoted to drawers and a fixed upper shelf, and the other to adjustable shelves. The case sits on a 3-in. high base (drawings left), with a 1½-in. toe space in the front and a 1½-in. overhang on each side. The base-frame height and the rear inset can be adjusted to clear any existing baseboard.

Start with the base, cutting the frame components to length and width for whatever size wardrobe you've decided to build. I rip sheet stock for the base, but you could also use 1x4s or even 2x4s.

Miter joints, which I use on the front corners of the base frame, may be made with an electric miter saw or a handsaw and a miter box. After the base pieces are cut to size, hold the front and back pieces together and mark on their top edges the location of the interior spacer at the middle of the base.

Finish nails and glue are adequate for assembling the cabinet base. You also may install glue blocks on the inside corners for additional strength. Biscuit joinery isn't necessary for the base because it is not subjected to much stress once the cabinet is installed, and using the joiner may actually take longer. But if you do use a biscuit joiner, be careful. The base pieces aren't wide enough to engage the anti-kickback pins on the faceplate of the joiner, so the tool may slip while cutting slots (for more on biscuit joinery, see *FHB* #70, pp. 50-53).

Cutting the components—Cut the case sides, the top and bottom pieces and the interior partition to length and width. Be sure the crosscuts are perfectly square, or the case may rack when assembled. I don't put a back on the case, so I use a cleat to help stiffen the case and to give me a way of attaching it to the wall of the closet. Cut a notch in the upper back corner of the center partition for a ¾-in. by 5½-in. cleat.

Shelf standards on the right-hand side of the case may be surface-mounted later. Or you can use a router and an edge guide at this point to cut shallow dadoes and let the shelf standards into the side of the case. But take care to stop the dadoes before they reach the top edge; otherwise, the ends of the dadoes will be visible after the case is assembled.

Lay out the location of the center partition on the top and bottom components. Also mark the location of the fixed shelf on the edge of the left side of the case and on the partition. Remember, however, the partition and the side pieces are unequal in height, which must be taken into account when marking the position of the fixed shelf. Also mark where top and bottom pieces will intersect the sides.

Before beginning the slotting process, dry assemble the case on a flat surface. You may need to apply light pressure with pipe clamps to stabilize the structure. When all case components are aligned to their marks, check to be sure everything has been cut to the proper length and that the case is square. Lightly mark the case pieces so that you can assemble them quickly and correctly later (drawing right). I chalk an "X" on the faces of the components to help locate the slotting jig and to orient pieces to the bench top during the end-slotting process.

Marking for assembly

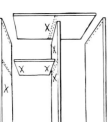

A simple and effective jig—The case is held together by 32 biscuits, which means that 64 slots had to be cut, and they had to line up with each other. To cut the slots quickly and accurately, I built a jig (drawing, p. 111) that locates the slots and helps orient and hold the biscuit joiner to the work. Positioned like a try square, the jig can

Closet wardrobe made easy. **A biscuit joiner and a layout jig make on-site construction of this space-saving closet wardrobe fast and reliable, which is an inducement for customers and an incentive for finish carpenters.**

be used from either the front edge or the back edge of the stock.

The entire jig can be made from ¾-in. hardwood plywood. It consists of an arm that is sandwiched between two crosspieces at a 90° angle. The crosspieces make up the base of the jig. The length of the arm extending beyond the edge of the base is the same as the width of the material you are joining—in this case 16 in. Once you have cut out the pieces for the jig, align them to a framing square and assemble them with glue and screws. But jigs can be made to any width and marked where you want to cut the slots. For the case of the wardrobe, I marked four evenly spaced slots for each joint.

Once the jig has been made, cut the slots in the side pieces of the case. These slots will receive the biscuits that are inserted into the ends of the top and bottom pieces. Position the jig back from the end of one of the case sides by the thickness of the stock and clamp the jig in place (top photo, facing page).

The centerline of the joiner's base should be aligned with one of the marks on the arm that indicates the center of a slot. Because the joiner is being used near the end of the workpiece, a scrap of case stock placed under the tool's face will add support and assure that the slot is perpendicular to the face of the work. This also is the time to slot the inside faces of the top and bottom pieces to receive biscuits that are inserted in the ends of the partition. Position the jig to the X side of the mark you made to locate the partition on the top and bottom pieces and make the slots. Slot the fixed shelf the same way.

Biscuit joinery gives you some latitude in the placement of the slots. In side-to-side alignment, discrepancies of up to ⅛ in. will not affect the outcome of the joint. But accurate vertical alignment is critical. Work away from the chip discharge of your biscuit joiner; otherwise, you will have to clear the work surface each time you reposition the tool for the next slot.

Cutting end slots—Use the same jig to position the joiner for slotting into the ends of the top, the bottom, the partition and the fixed shelf. For this operation, align the jig flush with the end of the piece to be slotted and clamp it securely in place (bottom photo, facing page). Orient the side of the piece marked with an X during dry assembly face down on the workbench and make sure the piece lies flat. The presence of chips or other foreign objects between the work and the bench, or warps in the bench itself, can easily misalign the end slots.

To slot the end of the stock, place the joiner flat on the work surface and orient the tool's centerline to one of the marks on the jig's cross arm. The base of the jig provides surface area for one of the joiner's anti-kickback pins, helping to hold the joiner in place. Again, work away from the expulsion of saw chips.

Assemble the case—When assembling a biscuit-joined case of this size and complexity, you must move quickly through the procedure. If the process goes on for more than about 15 minutes, some of the biscuits may swell from absorption

Making the drawers.
Drawers are made with the biscuit joiner and the jig shown on the facing page, with ¼-in. plywood bottoms and applied drawer faces of solid lumber. Make drawers 1-in narrower than the opening to leave room for the drawer slides.

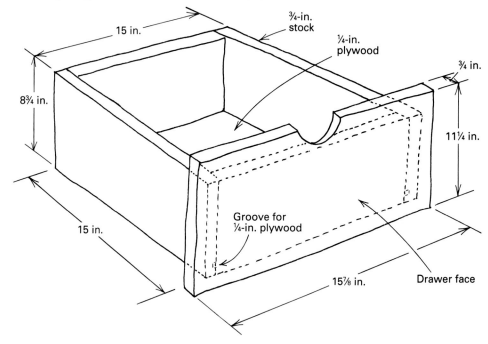

15 in.

¾-in. stock

¼-in. plywood

¾ in.

8¾ in.

11¼ in.

15 in.

Groove for ¼-in. plywood

15⅞ in.

Drawer face

of the glue, making adjustments impossible. If you are working alone, consider assembling this project in two stages.

The first stage is joining the center partition to the top and bottom pieces. Apply glue to the slots (I use yellow glue), preferably with a special applicator tip on your glue bottle to keep the process mess-free. Then insert the biscuits and pull this I-shaped assembly together with pipe clamps. Be sure the components stay perfectly square to one another. Don't move this assembly until the glue is completely dry (at least a couple of hours).

The second stage is joining the rest of the components to the first-stage assembly. Assemble the case on a flat surface that can support the entire structure. Orient the marked front edges upward, making sure all the components are in their proper positions. Then apply the glue, insert the biscuits and clamp the components together. Be sure the front edge of the boards are flush and take diagonal measurements to the outside corners of the case to check for square.

Instead of clamps, 1⅝-in. drywall screws may be used to hold the case together while the glue sets. Screw holes made through the sides of the case should be counterbored and then plugged. Or screws can be capped with plastic covers. These plastic covers are available through most mail-order hardware suppliers.

When the glue has dried, remove the clamps. Flip the case over on its face edge, then glue and screw the cleat to the notch in the partition. The ends of the cleat will butt against the inside faces of the case, where they are glued and nailed using finish nails. For additional strength, install several screws through the top into the cleat, counterboring and plugging the holes or capping the screws.

Making the drawers—The drawer components are cut, marked and joined just like the case was assembled (drawing above). Most drawer hardware requires a ½-in. space between the side of the drawer and the inside face of the case. Therefore, if the cabinet opening is 16 in., the drawers should be 15 in. wide. But check the specifications of the hardware you buy to see how much clearance the slides require. The drawer fronts are cut nearly as wide as the opening, which keeps the slide hardware out of sight when the drawers are closed. I use full-extension slides because they allow easy access to the entire depth of the drawer.

Cut the drawer parts to size (wait until later to cut the faces). Then dry assemble the drawers and mark them for location and orientation. You can use the same positioning jig to cut the slots for the drawers, but you will have to make new slot-reference marks because the drawer pieces are not as wide as the case. Cover the arm of the jig with masking tape, then mark the locations for three biscuit slots on the tape. Use the jig to make slots in the face and the ends of the pieces.

Cut a ¼-in. by ¼-in. groove along the bottom inside face of the parts to receive the ¼-in. plywood bottoms. Then glue and insert the biscuits, run some glue inside the groove and assemble the sides around the bottom panel. Drawer bottoms that fit snugly will help keep the drawer square when it is clamped. Apply band or pipe clamps, wipe off excess glue and check each drawer for square with diagonal measurements before setting it aside to dry.

Installing the hardware—Next, put the case on one side and lay out and install the shelf standards and the drawer slides. Flip the case over and install the hardware on the opposite side.

Secure the drawer hardware to the case sides using only the slotted holes; this will allow final adjustment later. A tick stick, a length of scrap lumber marked to show the location of each drawer slide, will help you lay out both sides of the drawer bay quickly and evenly.

Now it is time to install slides on the drawers. Cut the drawer faces and cut the semicircular hand grips in the top edges. Screw the faces to the drawers from the inside. If you use plywood for the drawer fronts, you will have to band the edges with a thin strip of solid material or veneer. Test the drawers by sliding them into the case on their slides. Adjust the gaps between the drawer faces by moving the slide hardware on the case sides up or down. When you are satisfied that the spaces between the drawers are even, secure the slides by adding screws to the round holes. Then cut the adjustable shelves to size and try them on the standards.

Installing the wardrobe—Set the base frame assembly, still unattached to the case, into position on the floor of the closet. Carefully level the frame from side to side and from front to back. Shim where necessary between the bottom of the frame and the floor. Screw the leveled frame to the floor by running 2-in. drywall screws from inside the frame down at an angle. Locate the screws so that they will pass through the shims, but be careful that screws don't break through the face of the base. Score and snap off the shims (or saw them off with a veneer saw) where they protrude from the frame.

Remove the drawers and the adjustable shelves from the case to lighten the load, then set the unit on its base frame. Be sure the overhang of the case is equal side to side and parallel to the face of the base frame. Then join the base and the case together. Finish nails will hold the case permanently in place.

Fasten the wardrobe to the wall by running 2½-in. drywall screws through the back cleat into wall studs. But make sure the face of the wardrobe is square before securing it to the wall. Because the case doesn't have a full back, it could rack during installation if you're not careful. If the wall is leaning back out of plumb, slide shims between the cleat and the wall surface first.

The hat shelf, with closet pole below, is built to suit the closet width and is best tended to after the rest of the wardrobe is in place. Cut the two shelf cleats, then drill holes in them to accept the ends of the 1⅝-in. closet pole. In locating the holes on the cleats, make sure a hanger will clear the shelf and the back wall. Or you can use surface-mounted rosettes for the pole, which are commonly available from building-supply stores.

Simply screw one cleat to the wall and another to the side of the cabinet with the pole in between (cut the pole about ³⁄₁₆ in. short to make its installation easier). Then cut the shelf to fit the opening, scribe the back and the far edge to the wall and drop it into place. ☐

Jim Tolpin is a woodworker and writer living in Port Townsend, Wash. Photos and drawings by the author.

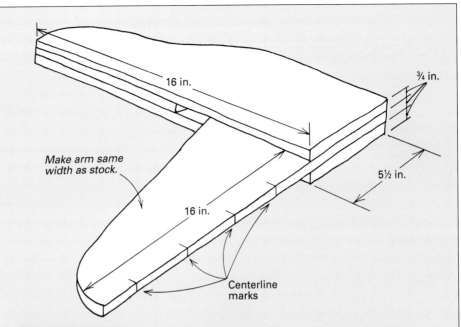

Biscuit-joiner jig.
This shop-made jig makes reliable biscuit joinery a speedy process. It can be used from either the right or the left and includes reference marks that line up with the centerline on the biscuit joiner.

Cutting side slots. With the jig lined up at the mark on the edge of the workpiece and clamped in place, the biscuit joiner cuts slots accurately. Note the scrap piece of case stock beneath the tool for added support. Here, the author uses a narrower jig than the one shown above to make an 11¼-in. deep cabinet. Jigs can be made to suit any cabinet width.

Cutting end slots. To cut the biscuit slots in the ends of the top and bottom pieces, the jig is positioned flush with the end of the workpiece. After the piece is clamped, make sure it is perfectly flat so that the slots will be aligned correctly. Then line up the joiner's centerline with marks on the jig and cut the slots.

Bed Alcove

Convert wasted attic space into a bed that has drawers, bookshelves and a vanity

By Tony Simmonds

When the middle one of my three daughters grew too old for the loft bed I built for her, the youngest, Genevieve, was happy to inherit it. The loft is in a small bedroom on the second floor of our house in Vancouver, B. C., Canada. Like many second floors of old houses, this one is really a half story, with sloped ceilings where the rafters cut across the intersection of wall and roof. The bedroom has only about 80 sq. ft., so its bed had to be on a raised platform to leave space for a dresser and a desk below.

Soon after she moved into the loft, however, Genevieve started bumping her head on the ceiling over the bed. When she eventually moved the mattress to the floor, I knew it was time for the old bed to go and for a new one to take its place. The bed alcove shown in the photo on the facing page was the result.

Will it fit?—The kneewalls that defined the sides of the room had originally been a little over 6 ft. high, leaving a great deal of wasted space behind them. I proposed to recover this space by moving the kneewall over 4 ft. to accommodate a 3-ft. wide mattress and a bedside shelf beyond that. Given the 12-in-12 pitch of the roof, this would bring the ceiling down below 3 ft. at the new kneewall. Would this be claustrophobic? To answer the question, I mocked up the space with packing crates and plywood to make sure there would be room to sit up in bed. A high ceiling is not a necessity over a bed—within reason, the reverse is true: A lower ceiling increases the sense of shelter and enhances the cavelike quality humans have always favored. Furthermore, a bed in an alcove that can be closed off from the rest of the room has qualities of privacy and quiet that are difficult to achieve in any other way. To get that extra layer of privacy, Genevieve and I decided that her bed alcove should have four sliding shoji screens.

The 9-ft. length of the space would provide room for a dresser and a vanity of some sort, as well as the bed. Drawers underneath the platform would triple the existing storage space. Light and ventilation would come from an operable skylight over the bed.

I had some misgivings about the location of this skylight in spite of the obvious benefits it would confer in terms of light and space. Having never slept directly under one myself, I didn't know whether a skylight so close to a bed would make sleep difficult. But in the end I was seduced by three arguments. First, the skylight would face north and therefore would not be subject to heat-gain problems; second, it would illuminate the shoji from behind; and third, there was the emotional pressure from my client—some drivel about the stars and the treetops and falling asleep to the sound of rain on the glass.

Tight layout—Juggling existing conditions is the challenge of remodeling. None can be considered in isolation. For example, I had to decide whether or not to keep the existing 7-in. high baseboard. I could have moved it, but I wanted to leave it in place, partly for continuity and partly to avoid as much refinishing as possible. Starting the drawers above the baseboard also meant that the baseboard heater already on the adjoining wall wouldn't have to be moved to provide clearance for the end drawer.

Four drawers fit into the space between the baseboard and the mattress platform. The drawers are 7 in. deep (6½ in. inside), which is ample for all but the bulkiest items. This brings the mattress platform to a height of about 18 in. With a 4-in. thick mattress on top of it, the bed still ends up at a comfortable sitting height.

In plan, the mattress takes up almost exactly three-quarters of the 9-ft. long space. The leftover corner accommodates a makeup table with mirror above and more drawers below. I imagined that the shojis would draw a discreet curtain over the wreckage of eyeliners, lipsticks, mousse and everything else that was supposed to go in the drawers but never would.

I knew that this vanity area, and especially the mirror, would need to be lit, but beyond making sure there was a wire up there somewhere, I didn't work out the details during the preliminary planning. I was in my fast-track frame of mind at this stage of the project.

Site-built cabinet—The underframe of the bed is a large, deep drawer cabinet. You could have it built by a custom shop while you get on with framing, wiring and drywalling. Custom cabinets are expensive, though, and after nearly 10 years in the business of building them, I appreciate the virtues of their old-fashioned predecessor, the model A, site-built version. It's economical in terms of material and expense, and you can usually get a closer fit to the available space.

The partitions supporting my daughter's bed are made from ⅜-in. plywood sheathing left over from a framing job (the rewards of parsimony). Each partition is made from three layers of sheathing (drawing below). The center layer runs the full height of the partition, but the outer ones are cut in two, with the drawer guide sandwiched between the top and bottom pieces. The guide is simply a piece of smooth, fairly hard wood, ¾ in. thick and wide enough so that it projects ⅜ in. into the drawer space.

Unless circumstances demand the use of mechanical drawer slides, I prefer to hang drawers on wooden guides. I have provoked derision from cabinetmakers because I use wooden guides in kitchens, but when it comes to bedrooms I am almost inflexible. Even large drawers like these will run smoothly year after year if they are properly fitted and if the guides are securely mounted. And for me there is a subtle but important difference between the sound and the feel of wood on wood vs. even the finest ball bearings.

I attach the guides with screws rather than with glue and nails so that they can be removed, planed and even replaced without difficulty should the need arise. A groove in the partition to house them is not necessary, but it's a way of ensuring that they all end up straight and exactly where you want them.

For this job, the pairs of guides on the three middle partitions had to be screwed to one another, right through the core plywood. I drilled and counterbored all the screws and clamped the partition to my workbench to make sure everything stayed tight while I drove the screws. Then, with the partition still on the bench and after inspecting every screw head carefully for depth below the surface, I set the power plane for the lightest possible cut and made three passes over each guide: first over the back third only, then over the back two-thirds and, finally, over the whole length of the guide. Tapering the guides so that they are a fraction farther apart in

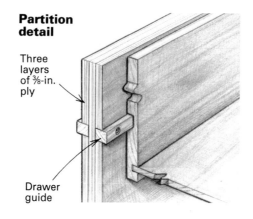

Partition detail

Three layers of ⅜-in. ply

Drawer guide

Tight fit. **Into this 9-ft. long space, the author squeezed a single bed, a row of 30-in. deep drawers, a bookshelf and a vanity. A recessed fluorescent fixture illuminates the mirror from above while the vanity table is lit by a lamp behind the mirror. The baseboard reveals the line of the original wall. Above it, drawer fronts cut from a single 1x10 are screwed from behind to the drawers. Photo by Charles Miller.**

the back allows the drawer to let go, rather than tighten up, as it slides home.

Partition alignment—Installing the partitions is the trickiest part of a site-built cabinet job like this one. I said earlier that you could save on materials by building the cabinet in place, but you can't save on time. After all, anyone with a table saw can build a square cabinet in the shop, but building one accurately in a closet or in an unfinished space under the rafters takes patience and thoroughness. The key to success is to establish a datum line, then lay out everything from this line, leaving the wedges of leftover space around the perimeters to be shimmed, trimmed, fudged and covered up as necessary.

In Genevieve's room, the existing baseboard provided a datum line in both horizontal and vertical planes. First, I divided the baseboard's length so that the four drawer fronts would lie directly below the shoji screens. I ran one screw into each supporting partition, about 1 in. below the top edge of the baseboard. Then I plumbed

the front edge of the partition and secured it with a second screw near the bottom of the baseboard. With the front edges located and the partitions standing straight, the next job was to align them to create parallel, square openings.

I built the new kneewalls 48¾ in. back from the inside face of the baseboard. This allowed me to run a couple of 1x4 straps horizontally across the studs to provide anchoring surfaces for the 48-in. partitions (drawing next page).

To align the partitions, I used hardboard cut to the full opening width (top left photo, p. 114). As long as the hardboard is cut square, and the partitions are secured so that the hardboard fits snugly between them, the resulting opening will also be square. I used screws to fasten the plywood flanges that held my partitions in place, just in case adjustment should be necessary.

When all the partitions were in place, I cut pieces of 1x2 to the exact dimension between each pair of drawer guides. Centered on the drawer fronts, the 1x2s are gauges that show how deep the grooves need to be in the drawer sides.

The bed slats also act as ties to link all the partitions together (bottom left photo, p. 114). I used dry 1x6 shelving pine for the slats, but almost anything that will span the distance between supports will do. I left an inch between the slats to keep the mattress well aired. I learned this the hard way when an early bed I built on a solid plywood platform developed mildew on the underside of the mattress cover.

Fitting the drawers—Before putting anything on top of the platform, I built and fitted the drawers. The drawers have ⅛-in. clearance between their sides and the partitions. The ⅜-in. projection of the drawer guide thus creates a ¼-in. interlock with the sides. All the drawers are 30 in. deep, but I let the sides extend 6 in. beyond the back of the drawer. The extensions support a drawer right up to the point where its back comes into view.

If time and budget allow, I use a router jig to dovetail the front of a drawer to its sides, but the back just has tongues cut on each end that are

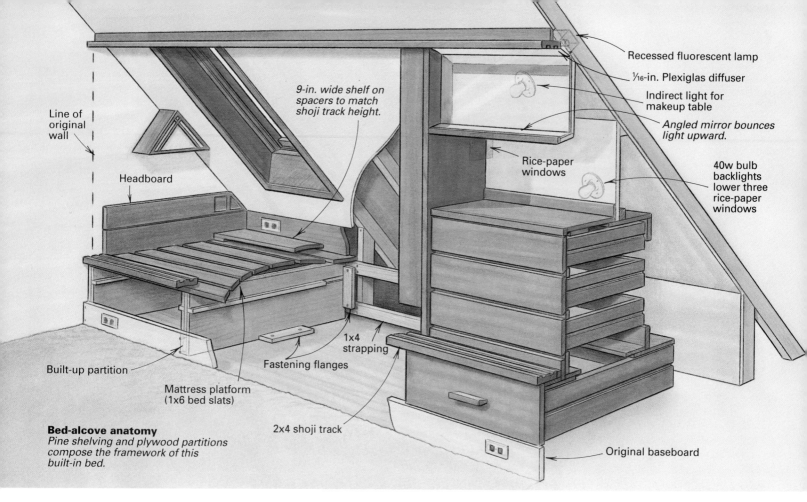

Recessed fluorescent lamp

¹⁄₁₆-in. Plexiglas diffuser

Indirect light for makeup table

Angled mirror bounces light upward.

40w bulb backlights lower three rice-paper windows

9-in. wide shelf on spacers to match shoji track height.

Line of original wall

Rice-paper windows

Headboard

Built-up partition

Mattress platform (1x6 bed slats)

Fastening flanges

1x4 strapping

2x4 shoji track

Original baseboard

Bed-alcove anatomy
Pine shelving and plywood partitions compose the framework of this built-in bed.

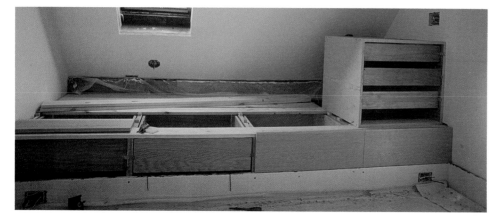

Aligning partitions. Load-bearing partitions made of three layers of ³⁄₈-in. plywood separate the drawer bays under the bed and support the mattress platform. The drawer guides are sandwiched between the outer layers of plywood. The photo at left shows the hardboard panels that helped to align the partitions. Once the panels were in place, the partitions were screwed first to the baseboard and then to strapping along the stud wall. The 1x2s clamped to the leading edges of the panels are gauges that will be used to determine the depth of the grooves in the drawer sides.

Linked by slat. The partitions are tied to one another across their tops by 1x6 pine slats (bottom left photo). Spaces between the slats provide ventilation for the mattress. At the right side, the carcase for the vanity drawers sits directly atop the bottom drawer partitions.

Bookcase wall. Shelves deep enough for paperbacks are affixed to a ³⁄₄-in. birch plywood panel between the bed and the vanity (photo below). The squares at the end of each shelf frame rice-paper windows that are backlit by bulbs behind the vanity drawers.

glued and nailed into dadoes in the sides (I take care not to put any nails where the groove for the guides will be plowed out). The drawer bottom rides freely in a groove cut in the front and the sides and is nailed into the bottom edge of the back, which is only as wide as the inside height of the drawer. Fastening the bottom here helps to keep the drawer square.

Fitting the drawers should present few problems if they are built square and true and if time and care have been invested in positioning the partitions. Don't try for too tight a fit, especially in the width of the groove. My guides were ¾-in. material, and I plowed out a ¹³⁄₁₆-in. dado in the drawer side. They're not sloppy.

On the other hand, you should be more stingy about the depth of the grooves. Remember, the guides have been planed to allow increasing clearance as the drawer slides home. Too much slop here can cause the drawer to bang about from side to side and actually hang up on the diagonal. You can always plow a groove out a little deeper. A router with a fence or a guide attached is the ideal tool for this because you can easily make very small adjustments. If things go wrong, you can glue a length of wood veneer tape into the dado, but it's nicer not to have to do that.

I dress the groove with paraffin wax, but only when I'm sure the drawer doesn't bind. Patience in working toward a fit has its reward here. The moment that a wood drawer on wood guides just slides into its opening and fetches up against its stop, expelling a little puff of air from the cabinet, is a moment that provides much satisfaction.

Beyond the footboard—With the drawers and the platform in, I had to decide what to do about the divider between the bed and the vanity. Here was where the self-imposed constraint of using the existing baseboard as the perimeter of the alcove began to bite. Because its height was determined by the slope of the ceiling, the mirror over the dressing table had to be as far forward as possible. But to bring it right up against the inside edge of the upper shoji track would eliminate the space required for a light above the mirror. And even that would put the top of the mirror at barely 6 ft. Temporarily derailed on the fast track, I tried to find other ways to light the mirror and kept coming back to the necessity of recessing a fluorescent fixture into the ceiling.

The fixture I used is a standard T-12 fluorescent fixture equipped with an Ultralume lamp (Philips Lighting Co., 200 Franklin Square Dr., Somerset, N. J. 08875; 908-563-3000). The lamp emits more lumens per watt than a standard cool-white lamp and has a higher Color Rendering Index, both important factors in getting an accurate reading on colors, like those at a makeup table.

Casting an even light across the face of the person standing at the mirror is important. So I put a

Headboard. A reading light inspired by the bookcase's square-and-triangle motif lights up the headboard side of the bed. On the left, wooden tracks for shoji screens frame the alcove.

narrow strip of mirror along the bottom edge of the large mirror, angled upward to bounce the light where it can fill in shadows.

Bookcase wall—As for the partition between dressing table and bed, my fast-track conviction that it could not be frame and drywall held up better. My daughter wanted more bookshelves, and the foot of the bed was a logical place to put them (photo facing page, lower right). I made the back of the bookcase out of ¾-in. birch plywood, which could be finished naturally on the book side and painted white on the dressing-table side to look like a wall.

To light the makeup table, I mounted a standard incandescent ceiling fixture in the space behind the mirror. On a playful impulse, I wired another of these lower on the sloped ceiling in the space behind the vanity drawer (the case for these has no back, so the fixture is easily accessible). Then, after carefully laying out the location of the bookshelf dividers and following a square-and-triangle motif suggested by the conjunction of the ceiling and the shelves, I jigsawed the holes in the birch ply and glued rice paper over them. This created little backlit rice-paper windows in the bookshelves. The dividers cover the edges of the paper. The only slight snag in this assembly is that the plywood thickness causes a shadow line, which can be seen where the backlighting travels at an angle through the window. If I'd thought of it in time, I could have easily eliminated the shadows by beveling these edges with a router.

The bedside reading light was more of a problem. Initially, I placed my standard ceiling fixture under the skylight as far down the slope of the ceiling as I could. I made a cardboard mock-up of the rice-paper shade that I had in mind to es-

tablish just how big it should be—the trade-offs being the height of the fixture, the size of the shade and its proximity to errant elbows. I thought I had a satisfactory balance, so I went ahead and made the lamp. But Genevieve put her elbow through it the first night she slept in the bed. I forgave her and accepted the lesson. The second reading light ended up above the head of the bed (photo left).

What about the shojis?—The shoji screens have yet to be made, and it now seems unlikely they ever will be. Although she was initially keen to have them, Genevieve now believes they would get in the way, and I agree with her. We analyzed the patterns of opening and closing that might be required during a typical day and night. It became clear that in spite of the desirability of drawing a curtain over the unmade bed by day and the unfinished homework by night, this teenager would rather live and sleep in one room—at least for the time being—than be bothered sliding screens to-and-fro all the time. A feeling of confinement was also a factor. Having tried out the bed myself one night when she was sleeping at a friend's house, I too felt I might want more distance between myself and any enclosing screen.

I admit that this was something of a blow to my vision of the room. What about the function of the skylight as a backlight for the shoji? What about the square-and-triangle motif I was going to incorporate into the shoji lattice? Ah, well, at least I hadn't made them already. And the grooves in the bottom track appear to work perfectly as 9-ft. long pencil trays.

The rejected shojis and the difficulties I had with the makeup light and the height of the mirror were all results of my decision to keep the bed alcove within the area beyond the existing kneewall. If I had moved this line 6 in. to 12 in. into the room, I could have raised the upper shoji track a few inches, creating plenty of space to mount the mirror light, the reading light and the shoji screens. The amount by which this would have reduced the size of the room would have been insignificant in relation to the space gained by building in the bed and the dressing table—a case of choosing the wrong existing condition to work from.

At least the client is satisfied. The project was completed during one of the long dry spells that Vancouver is famous for. Finally, one morning when the spider webs were glittering and the earth smelled refreshed and autumnal, Genevieve appeared downstairs with a beatific smile on her face. "It rained on my skylight last night," she said. □

Tony Simmonds is a designer and builder in Vancouver, B. C., Canada. Photos by the author except where noted.

Bookshelf Basics

A guide to support systems, shelving designs and materials

by Bruce Greenlaw

Bookshelves can transform a room into a library. Open shelves provide plenty of easily accessible room for books and collectibles. Glass-panel and raised-panel doors provide more secure shelf space, and related supplies can be stored in lockable drawers.

It's hard to define the quintessential bookshelf. The one above my writing desk, for example—a plastic-laminated particleboard shelf supported by three inexpensive metal wall brackets—was quick to build and perfectly suits my needs. The fixed shelf puts my reference books within arm's reach of my chair, and foam-padded steel bookends that I got at Wal-Mart keep them from falling off. Thos. Moser Cabinetmakers' furniture-grade cherry bookcases (Thos. Moser Cabinetmakers, P. O. Box 1237, Auburn, Maine 04211; 800-862-1973) (photo above), on the other hand, are designed to be heirlooms.

Between these two extremes lie a wide variety of shelving options. Basically, though, bookshelves are either housed in bookcases or supported by wall brackets (unless they're propped on milk crates or cinder blocks), and they're either fixed or adjustable. To my mind, the best shelving systems complement their surroundings and don't droop when they're loaded with books.

This article contains food for thought on designing bookshelves, plus an appraisal of shelving materials and hardware (sidebar p. 119). Of course, this information can be applied to virtually any type of shelving.

Photo this page: Courtesy of Thos. Moser Cabinetmakers

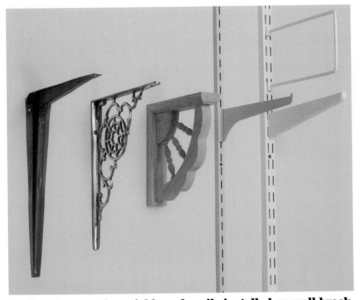

Bookshelves can be quickly and easily installed on wall brackets. Fixed brackets are screwed directly to the wall; adjustable brackets mount on slotted standards and can be moved up or down.

Shelf pins are inconspicuous. Shelf pins are less visible than shelf standards. Top to bottom: wire clip, which hides inside a groove at the end of a shelf; spring-loaded locking clip that prevents shelves from sliding or tipping; plastic clip; cushioned metal L-support; ornamental solid-brass pin; zinc-plated steel paddle; "library" pin with sleeve.

Measure the books, and size the shelves—Standard paperback books are about 4 in. wide (from the spine to the outside edge), but my binders, portfolios and biggest reference books are about 11 in. wide. Unless shelves will be used for storing old LP records (which are 12¼ in. wide) or giant art books, you'll rarely need to make bookshelves that are more than 11 in. deep. Some shops make 8-in. to 10-in. deep shelves and let wide books overhang.

If a series of shelves will be fixed in a bookcase or on nonadjustable wall brackets, measure the heights of the books that will be stored on them and space the shelves accordingly. Books range in height from about 6¾ in. tall for standard paperbacks to well over 12 in. tall, but most are in the 8 in. to 12 in. range. Don't forget to add ¾ in. to 1 in. of clearance above the books to allow fingers to grip them.

Shelf standards and hole-mounted clips make bookcases adjustable—Unless they're required for structural integrity, fixed shelves are probably unnecessary in bookcases. Adjustable shelves take much of the worry out of planning, offer more long-term flexibility and often result in more shelves fitting into a given space. The joinery for a bookcase without fixed shelves is simplified because you're basically just building a big box. Also, the supporting hardware for adjustable shelves is relatively inexpensive and easy to install.

Most shops don't use run-of-the-mill metal shelf standards to support exposed shelving. But metal standards are stronger than most alternatives, and I think they look okay if their color complements the surrounding bookcase. The standards are screwed or nailed in pairs to bookcase sides. They can be mounted on the surface, but most also can be recessed into dadoes for a more refined look (bottom photo). Barbed plastic standards that are simply pressed into dadoes (no nails or screws are required) also are available. But I've heard that these standards can be difficult to align, resulting in wobbly shelves.

Shelf pins are generally regarded as a step up from shelf standards (photo top right). These pins are strong enough for most applications, and they're less conspicuous than shelf standards. Shelf pins typically have ¼-in. dia. shanks that fit into holes of the same diameter. The holes are

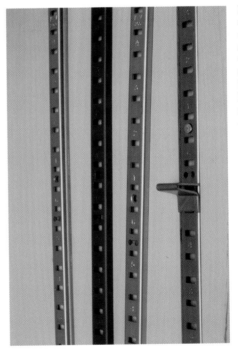

Metal standards are strong but conspicuous. Although metal shelf standards come with various disguising finishes, they tend to be more visible than hole-mounted shelf supports. Most standards can be recessed (at right) to make a finished appearance and to minimize gaps at shelf ends, or they can be surface-mounted.

spaced about 1 in. to 2½ in. o. c., and they're positioned so that shelves overhang the pins about 1 in. to 2 in. front and back.

Drilling shelf-pin holes can be tedious work, but a drilling template makes the job easier. A simple one can be made from a piece of tempered pegboard, which has ¼-in. dia. holes spaced 1 in. o. c. (top photo, p. 118). Using a drill bit with a drill stop mounted on it speeds the work and prevents the bit from boring through the workpiece.

Pegboard jigs don't last long, though. Commercial jigs are a better choice for production work. The acrylic shelf jig sold by The Woodworkers' Store (bottom photo, p. 118) has oversize holes that guide a self-centering ¼-in. Vix bit (or a 5mm Vix bit for metric work). The jig is only 19 in. long, but it

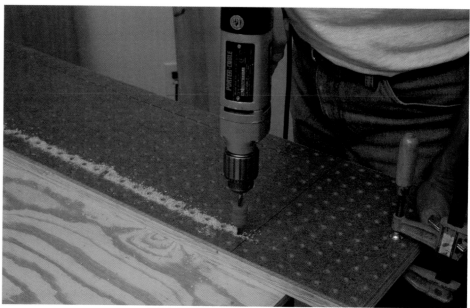

Templates simplify boring of shelf-pin holes. Drilling jigs make it easy to bore shelf-pin holes in parallel rows in bookcase sides. This disposable shop-made jig is tempered-hardboard pegboard, which has ¼-in. dia. holes spaced 1 in. o. c.

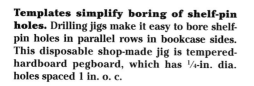

Commercial hole-drilling jigs are durable. The acrylic jig shown here (sold by The Woodworkers' Store; see sidebar, facing page) has oversize holes that guide a self-centering Vix bit. A shelf pin is used as an indexing pin to extend the range of the jig.

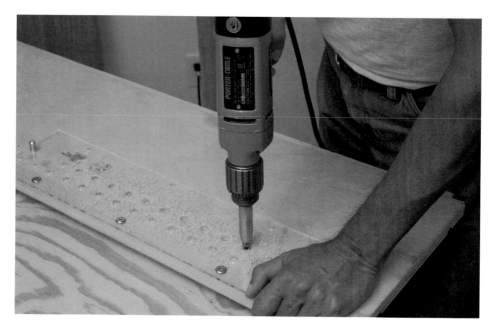

can be indexed with a shelf pin to bore any number of holes accurately. The jig costs $16, but the Vix bit adds another $35. Another shelf-drilling jig is made by Veritas tools and is available through mail-order outlets (Veritas Tools, 12 E. River St., Ogdensburg, N. Y. 13669; 800-667-2986).

Woodworking and hardware suppliers sell a variety of shelf pins and supports (photo top right, p. 117). Trussed plastic shelf pins are the least expensive and the most obtrusive. Metal L-supports are stronger and more subtle than plastic pins. Padded L-supports cushion glass shelves and help protect fragile finishes. Locking pins made of plastic or metal help prevent shelves from sliding and tipping. They allow bookcases to be shipped with shelves installed, and they help anchor shelves in earthquake country.

To my eye, the best-looking shelf supports on the market are spoon-shaped pins made of nickel or brass. Several shops I know of bore oversize holes and tap special metal sleeves into the holes to support these pins. The sleeves help prevent the holes from deforming, and they lend an air of refinement to unused holes. Dave Sanders & Company, Woodworker's Supply and others sell matching spoons and sleeves. Shelf pins can also be hand-carved, cut from dowels or even fashioned out of brass brazing rods. For invisible shelf support, wire clips are the best choice. They're inserted into holes bored into the sides of bookcases, and they hide inside sawkerfs cut into the ends of shelves.

Wall-mounted shelf brackets can be fixed or adjustable—Truth is, most of the bookshelves I've put up sit on plain metal wall brackets. The brackets install quickly with hollow-wall anchors or by screwing them directly to studs. Deluxe shelf brackets that install just as easily also are available. Two examples are the brass brackets sold by Renovator's Supply and the oak gingerbread brackets sold by The Woodworkers' Store (photo top left, p. 117).

For adjustable wall support, most hardware stores sell single-slotted metal standards that screw to walls and carry flimsy metal brackets that hook into the slots. But heavier-duty, twin-slotted systems (which have two rows of slots and hooks instead of one) are also available. The system sold by The Woodworkers' Store includes special screws that prevent shelves from sliding off the brackets and bookends that clip to the standards.

Dave Sanders & Company has the best selection of wall-mounted standards and brackets I know of. One type is mounted to studs before drywalling so that only a slim slot is visible afterward.

Well-designed shelves don't sag—Most cabinet shops use simple rules of thumb to determine shelf spans. Generally (when ¾-in. thick stock is used to support heavy reference books), particleboard and medium-density fiberboard (MDF) shelves span up to about 20 in.; softwood and plywood shelves span up to about 34 in.; and hardwood shelves span about 36 in. Shelves that carry paperbacks will span significantly farther.

For more predictable control over shelf sag, some shops use the shelf-deflection table published in Architectural Woodwork Quality Standards (available for $50, or $5 for members, from the Architectural Woodwork Institute, 13924 Braddock Road, Suite 100, Centreville, Va. 22020; 703-222-1100). The table lists the uniform loads that cause various unfixed 8-in. and 12-in. wide shelving materials to deflect ¼ in. when spanning 30 in., 36 in., 42 in. and 48 in.

AWI's chart points out a number of ways to beef up shelves to increase spans or to support unusually heavy loads. For instance, gluing a ⅛-in. thick edgeband to a ¾-in. thick particleboard shelf increases the shelf's load-bearing capacity by about 15%. Veneering the particleboard on two faces and one edge with ⁵⁄₁₀₀-in. thick plastic laminate increases its capacity by about 200%. Gluing a 1x2 solid-wood apron to the front edge boosts it by a whopping 300% to 400% (photo right). Greg Heuer, AWI's director of member services, notes that applying an apron to the back of a shelf, even pointing up (so that it hides behind books), gives the same results as a front apron. Although most people would consider a shelf deflection of ¼ in. to be excessive, halving the values listed in the chart would produce a less noticeable deflection of ⅛ in.

Several other strategies can be used for increasing shelf spans or load-bearing capacities. The most obvious is to use thicker shelves. Shelves made of 2x lumber, or two layers of ¾-in. plywood, for example, will span at least 48 in. Applying a half-round molding to the front edge of layered plywood creates a solid bullnose shelf.

For a significant span, consider building a torsion-box shelf, which works somewhat like a box beam or hollow-core door. These shelves consist of a grid of wood or plywood strips glued between two plywood skins. The biggest torsion-box shelf that I've heard of is 4 in. thick and 27 ft. long. Fixed shelves in bookcases can be reinforced by putting a back on the

Aprons stiffen the shelves. Putting a solid-wood apron on a shelf increases its load-carrying capacity. The sagging top shelf is ¾-in. particleboard. The virtually straight bottom shelf is ¾-in. particleboard with a 1x2 apron glued to the front edge.

case and fastening the back to the shelves, or by fastening intermediate support posts to the front edges. Adjustable bookcase shelves can be reinforced at the back with shelf standards and brackets or with hole-mounted shelf pins.

If you doubt the ability of a proposed shelf to support a given load, prop a sample shelf on a pair of blocks, load it with the weight it will carry and check the sag.

Recessing and trim make shelving look built in—Louis Mackall, owner of Breakfast Woodworks Inc. in Guilford, Connecticut, tells me that the best way to integrate shelving into a room is to recess it at least partway into a wall (top photo, p. 120), even if this procedure requires furring out the wall. According to Mackall, recessed shelves not only look better than

---

Sources for bookshelf hardware

Here's a short list of some useful companies to know about if you're in the market for bookshelf hardware.—*B. G.*

Dave Sanders & Company Inc.
107 Bowery, New York, N. Y. 10002
(212) 334-9898
Sells an impressive assortment of shelf standards, brackets and shelf pins, including Magic Wire concealed shelf supports and shelf pins with matching sleeves that reinforce pinholes.

Knape & Vogt Manufacturing Company
2700 Oak Industrial Drive NE, Grand Rapids, Mich. 49505-6083
(800) 253-1561
Makes cabinet and shelving hardware, including twin-slotted, wall-mounted standards and brackets, shelf pins and heavy-

duty metal wall brackets that support up to 1,000 lb. per pair.

Rangine Corporation
P. O. Box 128, Millis, Mass. 02054
(800) 826-6006
Makes storage systems, including the Rakks Shelving System: extruded-aluminum, wall-mounted shelf standards with locking, infinitely adjustable aluminum brackets.

Renovator's Supply
P. O. Box 2515, Dept. 9898, Conway, N. H. 03818-2515
(800) 659-0203
Sells ornamental brass shelf brackets.

Woodcraft Supply
P. O. Box 1686, Parkersburg, W. Va. 26102
(800) 225-1153

Sells assorted shelf pins, including solid-brass pins and brass sleeves that reinforce pinholes.

The Woodworkers' Store
4365 Willow Drive, Medina, Minn. 53470
(800) 279-4441
Sells a range of knockdown fasteners, standards, brackets and pins; the best assortment of edgeband I know of; and tools and accessories.

Woodworker's Supply Inc.
1108 N. Glenn Road, Casper, Wyo. 82601
(800) 645-9292
Sells assorted standards, brackets, shelf pins, and tools and supplies.

Recessed shelving doesn't intrude. Breakfast Woodworks in Guilford, Connecticut, recesses shelving to blend it with the architecture and to create the illusion of space.

Bracket-mounted bookshelves can look built in. Philadelphia woodworker Jack Larimore applies fancy aprons to basic wall-mounted bookshelves to produce ornate library storage that is available for a modest price.

projecting shelves, but they also appear to add space. Projecting shelves appear to subtract space. Continuing a room's base and crown moldings (if there are any) around a bookcase also helps to unite the bookcase visually with the room.

San Francisco woodworker Scott Wynn considers horizontal details in a room when sizing built-ins. If window head casings are a prominent feature, for instance, he'll make the built-ins the same height as the head casings. Wynn also makes bookcases supported by deeper base cabinets. The cabinet tops serve as oversize shelves that support art books. Wynn cautions that bookcase partitions must be placed directly over cabinet partitions or the shelves will sag.

Except for furniture-grade pieces such as Thos. Moser's, most bookcases are boxes with routed edges or applied trim. The boxes can have butt joints held together with nails or screws in concealed locations, or with biscuits or plugged screws in exposed locations (for more on biscuit joinery, see *FHB* #70, pp. 50-53). Special knockdown fasteners can also be used, allowing units to be dismantled and reassembled. Tall bookcases should have ¼-in. hardboard or plywood rabbeted into and tacked to the back for stability, a 1x nailer at the top for attachment to walls, or both.

Bookshelves don't have to nest in bookcases to look good. Philadelphia woodworker Jack Larimore has designed and built economical wall-mounted shelving that looks like pricey built-in furniture (photo left). The shelves are supported by metal shelf standards, but they're dressed up with decorative aprons. Ornamental metal and wood brackets can also be used to enhance the appearance of wall-mounted shelving (left photo, p. 117).

Top photo: Louis Mackall. Bottom photo: Mitch Mandel/Rodale Stock Images.

Solid wood is beautiful but unstable—Thos. Moser builds most of its bookcases out of ¾-in. thick American black cherry. At the other extreme, I've used low-cost #3 knotty pine for a number of shelving jobs because it's relatively strong and inexpensive. Unlike some shelving materials, solid wood doesn't require edgeband, and edges can be detailed easily with a router or a hand plane.

Unfortunately, solid wood is unstable. It shrinks as it dries, and it expands and contracts across the grain in response to changes in humidity. It also tends to warp and cup more than sheet stock, causing adjustable shelves to seesaw.

One nifty idea I've seen is to use bullnose hardwood stair-tread stock for shelves. It usually comes in red or white oak, it's often glued up out of narrow strips for dimensional stability, and it's a sturdy 1 in. to 1¹⁄₁₆ in. thick. Millworks and many lumberyards sell it, and when you factor in presanded finish and ready-made edge treatment, the price is reasonable.

Hardwood plywood is an industry favorite—Most shops I spoke with shun solid wood in favor of sheet stock such as hardwood plywood, particleboard or MDF (bottom photo). Sheet stock is more stable than solid wood, and it normally can be turned into shelving parts faster because it doesn't have to be flattened and straightened as solid wood often does.

Hardwood plywood is available with a veneer core, an MDF core, a particleboard core or a lumber core. Veneer-core plywood is strong, has excellent dimensional stability, holds screws well and is widely available (especially in birch, red oak and lauan). But it can have voids in the core that could occasionally cause a hole-mounted shelf pin to sag or to fall out.

MDF-core and particleboard-core plywood don't have voids, but they also don't hold screws as well as veneer-core plywoods do, although using particleboard screws (which have a deep, coarse thread) helps. These panels are also heavy, weighing almost 100 lb. per ¾-in. thick sheet. This weight not only makes them awkward to machine, but it also makes them a questionable choice for portable bookcases.

Nevertheless, some shops use MDF-core plywood exclusively for their bookshelves. They like its solid core and its cost, which is about 5% to 20% less than veneer-core plywood. Georgia-Pacific's Fiber-Ply core plywood and Columbia Forest Products' Classic Core plywood are compelling hybrids. They look like conventional veneer-core plywood, but they have homogenous layers of fiberboard instead of solid-wood plies directly beneath the face veneers.

Lumber-core plywood is strong, it holds screws tenaciously, and it is supposed to be voidless. But today's lumber-core plywood, much of which is imported, can have shrinkage voids in the core. Good-quality lumber-core plywood is expensive and can be hard to find.

Regardless of the type of plywood used, exposed edges need to be covered with moldings or edgeband (top photo). Moldings can be anything from pine screen bead to solid-wood or plastic T-moldings that fit into grooves in panel edges. Edgeband is peel-and-stick, iron-on or glue-on veneer that comes on a roll. Solid-wood, polyester and even metallic-foil edgebands are available.

Particleboard and MDF work well for paint-grade work—Particleboard and MDF are hard to beat for paint-grade shelving systems. They're relatively inexpensive and stable, and they don't require edgeband. Most lumberyards sell ¾-in. industrial-particleboard shelving with filled bullnose edges. Available in many widths—typically 8 in. or 12 in.—precut shelving is quick and convenient.

Called the "Buick of particleboard" by Albuquerque woodworker Sven Hanson, MDF is made of highly compressed wood fibers instead of parti-

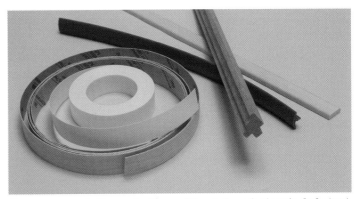

Cover exposed plywood edges with edging. Options include (top) T-molding; pine screen bead; plastic T-molding; peel-and-stick hardwood edgeband (outer roll); and iron-on polyester edgeband (inner roll).

Sheet stock is a stable and uniform choice for shelving. Options range from hardwood-veneer plywood to medium-density fiberboard panels with a durable melamine surface.

cles, resulting in panel surfaces that are as flat as glass and edges that machine beautifully without chip-out (machining produces clouds of fine dust, though, so wearing a high-quality dust mask is a must). The Home Depot in my area sells ¾-in. MDF for about $20 per sheet, which is a bargain.

One problem with MDF and particleboard panels is that unsealed edges drink paint, resulting in an uneven finish. You can seal MDF edges with two coats of water-based polyurethane finish, but quick-drying sanding sealers and white PVA glues diluted 20% with water will also work.

Paint-grade plywood, called MDO (medium-density overlay), can also be used for painted shelving. This exterior-grade veneer-core plywood is coated with a resin-treated paper that makes a superb, smooth substrate for paint. MDO can also be taped like drywall, so it makes a good material for built-in shelving.

MDF and particleboard can also come veneered with melamine or plastic laminate. Melamine is a thermally fused, resin-impregnated sheeting that resists abrasion, stains, heat and chemicals, and it's available in a variety of sizes, patterns and colors. The Home Depot sells melamine panels that are ripped to standard bookcase widths and that are predrilled with ¼-in. holes that accept standard shelf pins. Melamine can be tricky to work with because it chips easily. But it's a versatile material, and it has a hard prefinished surface. (For more on working with melamine, see *FHB* #99, pp. 68-73). □

Bruce Greenlaw is a contributing editor of Fine Homebuilding. *Photos by the author, except where noted.*

A Home Library

Maple, purpleheart and an enthusiasm for small details

by Charles Wardell

A writer once said that there's no furniture as charming as books. Too often, though, the furniture made for books barely rises above the level of utility. But high-grade materials, a thoughtful design and attention to detail can turn what might have been little more than storage into a home's focal point.

Such was the case when Robert Hillen of Interior Woodworking in Cincinnati was hired to redesign the interior of a condo. The new owners were trading down from a larger home, and though that meant shedding excess possessions, there was one thing they were unwilling to part with: their 360 lineal feet of books. In fact, they wanted room to expand it to 420 ft. Hillen's design included a complete redo of the interior, but its centerpiece was a massive maple bookcase that covered 30 ft. of the living room wall (photo right).

Durable elegance—After completing the design, Hillen handed the project over to the company's head woodworker, Bradd MacCallum. MacCallum, who spends the bulk of his time in the shop building custom furniture, has a head for numbers and a passion for accuracy. Everything was built in the shop, then assembled like a puzzle at the job site. This demanded both a great deal of planning and precise measurements. Many of the pieces were made to tolerances of $\frac{1}{64}$ in.

Although the bookcase would be a dominant feature of the house, it couldn't be overwhelming. Purpleheart detailing was used all around to this end, helping to make the bookcase an engaging piece of workmanship, rather than an imposing mass of wood. The purpleheart's deep red contrasts sharply with the blond maple of the shelves, while at the same time complementing the violet hue of the carpet. The details are where the furnituremaker's hand really shows, enticing the observer to draw closer rather than forcing him to step back.

But if appearance was important, the weight of all those books meant that strength was crucial. Luckily, the condo had been built on a thick concrete slab, so floor strength wasn't

When Robert Hillen was asked to redesign the interior of a condo, the main question was what to do with all the clients' books. Hillen answered with an attractive wall unit that covered the 30-ft. length of the living room wall.

as much of a problem as the strength of the bookcase itself. Hillen specified a 5-in. thickness for the two center uprights, and 3 in. for the rest. Solid wood was out, though. Not only would solid maple be unreasonably expensive, but its weight would make the uprights too heavy to transport and install. MacCallum considered using wooden frames clad with ¾-in plywood, but those would present similar problems. Finally, he hit upon a solution.

A high-grade prefab—To make the uprights, MacCallum built a sandwich of Styrofoam and ½-in. bird's-eye maple plywood (see drawing on p. 124), laminated together with Wilsonart spray epoxy (Ralph Wilson Plastics Co., 600 General Bruce Dr., Temple, Tex. 76504; 817-778-2711). The sandwich is both lighter and less costly than framed uprights would have been, so MacCallum could install them with one helper. The foam added stiffness to the assembly, enabling him to use ½-in. plywood for the sides instead of ¾ in.—with a 30% savings in cost. The system left the option of easily trimming the uprights if need be, because the completed sandwich could easily be run through a table saw. And Styrofoam muffles sound; by contrast, a wooden frame with plywood sheathing would feel and sound hollow.

Half-round maple facings flanked by purpleheart accent strips cover the leading edge of the uprights. The facings were made from glued-up maple that was rounded with a router and a carbide quarter-round bit, then glued to a flat maple base (photo on p. 124).

To ensure a tight joint at the wall, MacCallum scribed and trimmed the backs of the uprights. Then, using a router and template, he removed the foam at the top, bottom and rear edges of each upright to a depth of 2 in. The resulting channel at the rear slid over 8/4 by 2-in. poplar nailers that had been screwed through the finished drywall to horizontal blocking between the wall studs. Finish nails secure the panels to the nailers.

A router was also used to make the holes for the brass shelf pins (Hafele American Co., 3901 Cheyenne Dr., Archdale, N. C. 27263; 800-334-1873). MacCallum made an 8-ft. long birch plywood template with ⅝-in. dia. holes at each shelf-pin location. After clamping the template to an upright, a plunge router with a ⅝-in. bushing and a ¼-in. straight plunge-cutting bit made short work of the holes. There were advantages to this method, the most important being that it proved exceedingly accurate. The router also did a cleaner job than a drill; there was no tearout at any of the 1,000 or so holes. And sharpening the bit made it a hair undersize, just small enough to guarantee a secure fit for the ¼-in. shelf pins.

Of course, aligning the finished uprights during installation was essential, if the shelves themselves were to line up. A level line snapped on the wall and a corresponding mark on the back edge of each upright solved that problem. The bases of the uprights rest on poplar blocks with lengthwise shoulders (drawing on p. 124) that were first screwed to the

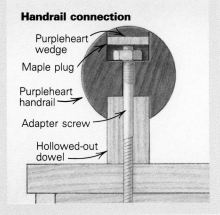

The purpleheart handrail doubles as a support rail for the rolling ladder (photo above). The ladder's two pairs of rail hooks allow it to be hung from the rail vertically for storage or slid along in a diagonal position for use. Note the felt pads on the underside of the hooks; they help the ladder slide along the railing.

Handrail connection

Purpleheart wedge

Maple plug

Purpleheart handrail

Adapter screw

Hollowed-out dowel

The ladder wheels were made in Hillen's shop after he tried—and failed—to find an off-the-shelf set that worked with the design.

concrete floor with Tapcon screws (ITW Buildex, General Construction Systems, 1349 W. Bryn Mawr Ave., Itasca, Il. 60143; 708-595-3500). Each upright was shimmed until the mark on the back met the line on the wall. The block and the shim space were later covered with baseboard.

Holding the half-round upright facings back 6 in. from the floor left room for a purpleheart plinth block. MacCallum cut these blocks a bit

long, then trimmed them on site for a tight fit to let the front of each unit sit squarely on the floor, in spite of a varying floor level.

A 4/4 poplar block glued into the top of each upright served as a nailer for the purpleheart top shelf and built-up pediment. MacCallum made the top shelf oversize, scribed its back edge to compensate for variations in the wall plane, then cut the leading edge. This gave him a tight joint at the wall and a consistent ⅝-in. overhang in front. The pediment was crowned with a purpleheart arch, added on site to achieve a precise fit (top photo on p. 124).

The shelves themselves are 5/4 maple joined edge to edge. They're trimmed with a half-round purpleheart accent strip set in a routed slot at each shelf's leading edge. At the back, a ¼-in. gap separates the shelves from the finished drywall. This gap not only protects the wall, but it leaves room for wiring, giving the clients the option of using the shelves for speakers or other electronic equipment.

A double-duty railing—One of the most ingenious parts of the bookcase design is the continuous purpleheart support rail for the rolling library ladder (top photo, left). After taking a 20-ft. run across the front of the bookcase, it turns a corner, goes across the balcony, then travels down the U-shaped staircase. The railing was made from glued-up strips (3 laminations in all). Their edges were rounded with a quarter-round carbide bit. Routing one edge at a time, MacCallum was able to cut about ¹⁄₁₆ in. per pass. All joints are staggered at least 2 ft. apart. A sliding table rig on the shop table saw ensured precise crosscuts. Although the railing itself was laminated with yellow woodworker's glue, the turns were joined to the straight runs with dowels and cold-pressed resin glue (National Casein Co., 601 W. 80th St., Chicago, Il. 60620; 312-846-7300). It was difficult to clamp the curved connection accurately; the resin glue set up fast enough to let the connection be made with hand pressure.

The purpleheart rail is supported by adapter screws—double-ended screw-bolts with machine and wood threads—covered with hollow dowel spacers that MacCallum bored out in the shop with a router (drawing left). The wood threads were turned into the upright facings, the balcony and the stair half-walls; the dowel was then installed over the machine threads and the railing was fitted over the dowel. A countersunk hole at the front of the railing allowed the installation of a nut over the machine threads. This hole was later filled with a slotted maple plug. A purpleheart wedge forced into the slot secured the plug in the hole. The whole assembly—shelves, uprights, railings—was sanded smooth with a palm sander and finished with three coats of Watco oil.

The library ladder—A rolling ladder is a welcome touch to any library, but Hillen's 16-ft. high shelves would have been useless without one. The ladder rails taper from a width of 3½ in. at the top to 5 in. at the bottom. The steps sit in dadoed slots and are held fast with #8

The bookcase was capped with a built-up pediment, fabricated with foam and plywood just as the shelf uprights were.

Purpleheart details were used all around. They encourage the observer to look closer.

wood screws 2½ in. long. The countersunk holes were plugged just like those in the railing, but this time with purpleheart plugs and maple wedges. The steps are 4 in. deep. Like the shelves, they are trimmed with half-round purpleheart accents set into their leading edges. A 6-in. shelf at the top of the ladder serves as a seat for browsing.

The top of the ladder sports two pairs of maple rail hooks (top photo). Each was made from 20 pieces of maple veneer laminated around a curve that matched the handrail. Their undersides are covered with felt so that they'll easily slide along the rail. The top hooks are the working hooks; when the ladder is in use, they hold its wheels 3 ft. from the front of the bookcase. And while the angle of the ladder is admittedly on the steep side, it was the best that the available floor space would allow. The lower pair of hooks allow the ladder to be hung vertically, keeping it out of the way when floor space is needed for other activities.

Hillen had originally specified stock wheels for the ladder. Unable to find any that worked well with the design, however, he asked MacCallum to make a pair (bottom photo on p. 123). Each wheel consists of two pieces of cross-laminated maple that turn on a $\frac{5}{16}$-in. metal axle and a pair of flanged roller bearings. The wheel assemblies are set into slots at the base of the ladder. □

Charles Wardell is an assistant editor of Fine Homebuilding. *Photos by the author.*

Drawings: Bob LaPointe

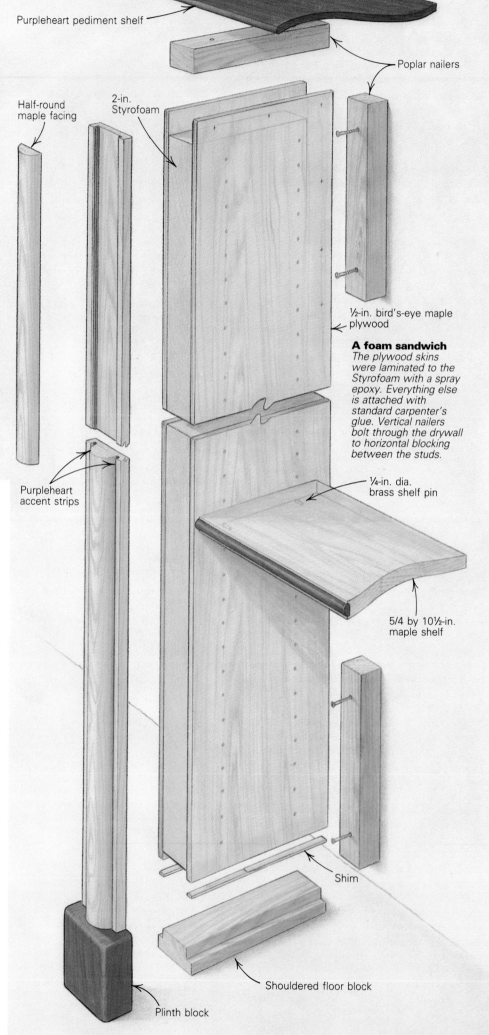

Purpleheart pediment shelf

Poplar nailers

Half-round maple facing

2-in. Styrofoam

½-in. bird's-eye maple plywood

A foam sandwich
The plywood skins were laminated to the Styrofoam with a spray epoxy. Everything else is attached with standard carpenter's glue. Vertical nailers bolt through the drywall to horizontal blocking between the studs.

Purpleheart accent strips

¼-in. dia. brass shelf pin

5/4 by 10½-in. maple shelf

Shim

Shouldered floor block

Plinth block

INDEX

The articles in this book originally appeared in *Fine Homebuilding* magazine. The date of first publication, issue number and page numbers for each article are given at right.